This Is Server Country

THIS IS SERVER COUNTRY

AI, Power, and the Remaking of Rural America

MICHAEL J BOMMARITO II

FIRST EDITION

2026

This Is Server Country:
AI, Power, and the Remaking of Rural America

Cover design and typesetting by the author. Printed in the United States of America.

Author Website: `michaelbommarito.com`

ISBN 979-8-9943457-3-3 (paperback)

ISBN 979-8-9943457-2-6 (ebook)

Publisher's Cataloging-in-Publication Data
Names: Bommarito, Michael J., II, author.
Title: This is server country :
   AI, power, and the remaking of rural America / Michael J Bommarito II.
Description: First edition. | Michigan : 2026. | Includes bibliographical references.
Identifiers: ISBN 979-8-9943457-3-3 (paperback) | ISBN 979-8-9943457-2-6 (ebook)
Subjects: LCSH:
   Data centers—United States.
   Artificial intelligence—Economic aspects—United States.
   Electric power—United States.
   Infrastructure (Economics)—United States.
   Rural development—United States.

Classification: LCC TK5105.85 .B66 2026 | DDC 004.6—dc23

To our children:
may you live a better life for these decisions we make.

Author's Note

I am writing this book in late January of 2026. By the time you read it, the facts will have changed, but I wager the issues will remain.

Over the last seven months, I have documented more than 600 data center projects across all fifty states. Together they represent over a trillion dollars in announced investment and 132 gigawatts of planned power capacity. The database captures the eight largest hyperscalers driving demand, the sixteen major operators pouring concrete, the eleven primary financial sponsors structuring the deals, and the many utilities caught in the middle.

Some facilities I have tracked are operational. Others exist only as construction sites and permits. Many will never pour a yard of concrete. Thankfully, I have accepted the inevitability of a second edition; the very nature of this book will no doubt require updates.

Even in the last thirty days, a new 345 kV transmission line has been proposed that would run just a few hundred yards from my century-old bank barn. The feverish urgency with which I have written thus needs little explanation.

Beneath the media leaks and eddies of speculation, at least the hard anchors hold: township committee minutes, SEC filings, utility regulator dockets. These public records are the foundation of this book.

We have been here before. The railroads opened the West. The interstates created the suburbs. Each time, we built the future with concrete and steel. Each time, benefits flowed broadly while costs concentrated locally. The AI buildout follows the same pattern, at comparable scale, at faster speed.

But a database is not a story. To understand the machine, you have to stand in one place and watch it work.

In October 2025, Governor Whitmer announced that OpenAI and Oracle would build a $7 billion campus in Saline Township, Michigan. Saline is not the largest project in my database, nor the most controversial. But it is representative. The same forces acting on this Michigan township—power constraints, state competition, community resistance—are reshaping Virginia, Texas, and Arizona. This is a book about America, not just Michigan.

Michigan also happens to be the place I know best.

———

My wife and I grew up in mid-Michigan, returned when we started a family, and now live an hour from where this story begins. We earned five degrees between us at the University of Michigan, fifteen miles from the Saline site. We have a small farmstead. I have read the *Michigan Farm News* at home for years; that is where I first saw these issues stirring, long before they appeared in my FT subscription.

I also understand the other perspectives. I lead the ALEA Institute, a nonprofit focused on responsible AI—training open models, releasing open data, conducting policy research. I was part of the research team that tested GPT-4 against the bar exam. I even worked, once upon a time, in Manhattan for hedge funds.

More pointedly, I used AI to help write this book. Claude Code, Anthropic's agentic command line tools, assisted with research, drafting, and revision. Every query consumed compute cycles in data centers like the ones I describe. The irony is not lost on me. I am complicit in what I document.

I am no main character in this story, but I hope you trust that I speak from experience as well as conviction.

———————

Most Americans will never notice a data center. The buildings are designed to be invisible: windowless boxes sheltered behind security fences on land that used to be something else. The electricity they consume shows up as flickering lights or rising utility bills no one can trace. The decisions happen in boardrooms and regulatory dockets and township meetings that rarely make the news.

I called this book *This Is Server Country* because the phrase echoes something you see on roadside signs across rural America: *This is God's Country, This is Trump Country*—declarations of identity and providence.

The AI buildout is making a similar claim, written in concrete and copper instead of paint and prayer. Whether that transformation is progress or loss depends on who you ask. I have tried to hold both futures in superposition together. Those who know me may suspect I hold opinions stronger than those expressed here. They would be right.

One question in particular runs through this book: why farmland? America has thousands of shuttered factories, abandoned malls, and decommissioned power plants—sites already scarred by industry, already connected to roads and utilities. Why does the AI buildout choose virgin soil instead? I came to this project assuming the answer was greed or indifference. The truth turned out to be more structural, more frustrating, and more worth understanding.

But I have done my best to tell the story, the whole story, and nothing but the story, quantum though it may be.

That is harder than it sounds. Revealing all perspectives in their fullness requires more words than most would read. An earlier draft of this book swelled to nearly 550 pages, dense with footnotes and academic hedging. I hope the reader forgives the fences I have thus strung around the boundaries of this book.

———

Some things I could document. Some things I had to imagine.

Some characters in this book are composites—fictional people representing documented roles: *Frank, Ellen, Harold, David, Steve, Roberto, Sarah,* and others. Their dialogue and inner thoughts are invented, though grounded in trade publications, public testimony, and news accounts. You will recognize these composites by their italicized single names. Where real people appear by full name—governors, executives, regulators—their words come from the public record.

———

This is a book about AI, power, and politics: the silicon that demands electricity, the land that hosts it, and the deals—from township zoning to trade policy—that determine where it gets built. It does not address what AI might eventually become—whether these systems will remain tools or become something else entirely. That question deserves its own book.

It does not argue that data centers are good or bad. It argues that they are consequential. The buildings going up now will stand for decades. The communities hosting them will be transformed whether they chose transformation or not. The farmland beneath them will not return.

The choices being made now deserve more scrutiny than they are getting. This book is my attempt to pay attention before the concrete sets and the soil is gone.

—*MJB, January 2026*

Contents

The Token

YOU TYPE A QUESTION. A second later, words appear. The cursor blinks. More words arrive, assembling themselves into sentences, paragraphs, thoughts. You barely notice the miracle.

It is a Tuesday evening in January 2026. You are sitting in a coffee shop on Liberty Street in Ann Arbor, Michigan. The winter dark has already settled in. Through the window, students trudge past in heavy coats, their breath visible in the cold. Your laptop glows on the table beside a cooling cup of coffee.

The response appears at a pace that mimics typing. But no one is typing. No human sits on the other end of this conversation. The words emerge from a process so complex, so distributed, that even the engineers who built it cannot fully explain how it works.

Between your fingertips and that answer, a vast infrastructure hums. Fiber optic cables carry your question at the speed of light. Servers parse your words into tokens—numerical fragments that neural networks can process. And somewhere, in a building most people will never see, electricity flows through silicon at rates that would have astonished the engineers who built America's first power plants.

Fifteen miles southwest of where you sit, the answer takes physical form. A facility nicknamed "The Barn" is rising across 250 acres of former farmland in Saline Township. Three buildings, each the size of five football fields. When fully operational, this complex will consume 1.4

1

gigawatts of electricity—roughly the output of a large nuclear reactor. Enough to power a city of 800,000 people.

Your question consumes a sliver of electricity. A few watt-hours, perhaps a penny's worth of power. Trivial. But you are one of hundreds of millions asking questions today. And that chat window is only what you see. AI inference runs beneath the surface of modern life. It shapes the search results that now include AI summaries. It autocompletes your emails. It decides what appears in your social media feed and recommends what you watch next. A typical American triggers thirty to a hundred AI inferences per day, often without knowing it. Power users, developers with coding assistants and professionals with AI-augmented workflows, may trigger hundreds. The variation is enormous, the accounting imprecise. No one tracks the total. And that is just individuals; enterprises and governments run millions more through APIs that never touch a chat interface.

Multiply these invisible queries across a population, and the slivers become a torrent. ChatGPT alone processes 2.5 billion prompts daily— and ChatGPT is one product among dozens. The infrastructure exists not for any single question but for the staggering volume of them, every hour of every day, from every corner of the world. Seven billion dollars in construction. 1.4 gigawatts of continuous power. All to serve the collective appetite for answers.

The token that appears on your screen, each fragment of a word, represents electricity transmuted into something that resembles thought. Every token is cognitive labor, produced at industrial scale, at a cost no human worker could match. The infrastructure rising across America is not just serving curious students. It is converting capital into cognition—and reshaping the economy in ways we are only beginning to understand.

Saline Township sits in Washtenaw County, about fifteen miles southwest of Ann Arbor. Farm country. Flat fields and two-lane roads, scattered houses on large lots, 2,300 people spread across nearly thirty-five square miles. Downtown has a veterans' memorial in Oakwood Cemetery and City Limits Diner, where farmers have had coffee in the same booths for forty years. Henry Ford rebuilt the dam in 1936 to power a soybean plant that bought crops from hundreds of local farmers. The plant closed decades ago. The dam remains.

The township government is modest: a five-member board, a small budget, a volunteer fire department. Until 2025, the biggest controversies involved zoning variances for home additions. Board meetings attracted a handful of regulars. Decisions were made without fanfare.

Frank had been township supervisor for eight years. A retired auto industry engineer, he ran for the board after getting annoyed about a drainage issue near his property. The job paid a modest stipend and consumed evenings that might otherwise go to his grandchildren—*Lily*, eight, and *Marcus*, six, who lived twenty minutes away in Ann Arbor and asked every Sunday why Grandpa couldn't come to soccer. His wife asked the same question, in different words. But he took the job seriously. Someone had to.

In January 2025, nothing in his experience prepared him for what came next.

———

In early 2025, a developer called Related Digital began quietly assembling parcels of farmland along Michigan Avenue. The road has another name: US-12, the old Chicago Road. Pioneers cut it through the wilderness in 1827, following Indian trails west. For a century it was the main route between Detroit and Chicago, and Saline grew up along it, serving travelers and farmers heading toward a city that barely existed.

Then I-94 opened in 1961. The interstate rendered the old road a relic. Some farms along it have been in the same families for over a hundred years. Metal plaques—some donated by the local utility—mark the survivors: Centennial Farms.

That developer assembling parcels along Michigan Avenue was not a local operation. Related Digital is the data center arm of Related Companies, founded by Stephen Ross—the billionaire who owns the Miami Dolphins and whose name adorns the business school at the University of Michigan, fifteen miles up the road in Ann Arbor, and whose ties to the state run deeper still. The people who built Hudson Yards in Manhattan had come to Saline to capitalize on the AI boom.

The company filed for rezoning without revealing who would occupy the site. Residents noticed the activity, and questions arose at board meetings. Representatives spoke of "advanced technology" and "high-wage jobs" but deflected specifics. Who was the tenant? They could not say. How much power would the facility consume? "Significant." What about water usage, traffic impacts, noise levels? Vague assurances.

The Planning Commission had already rejected the proposal in August. Now, on September 10, 2025, the full Board would vote. Sixty residents packed the Saline Township hall—more than *Frank* had ever seen at a board meeting, though fewer than the developer had probably expected. Outside, the evening had turned crisp, the kind of September night that reminded you summer was over. Inside, the room grew warm with bodies and voices.

The public had questions. Traffic. Water. Pollution. Sound. What happens when the power grid fails? What happens when the company sells? *Ellen*, whose family had farmed adjacent land since 1947, spoke about groundwater concerns. She was sixty-four years old, had taught biology at Saline High School for thirty years before retiring, and could identify every bird that crossed her property by its song. Her hands shook as she read from handwritten notes. "My grandfather dug that well with his own hands," she said. "Fifty-two feet deep, through clay and sand, with a shovel and a mule-drawn pulley. You're asking us to trust that a company we've never heard of will monitor it, fix it if it runs dry, and honor that promise twenty years from now. What happens when they sell to someone else?"

The developer's attorney responded with slides and statistics. Projected employment: 450 permanent positions. Tax revenue increases: 400 percent. Environmental monitoring: comprehensive. *Frank* found himself wondering how many township boards this lawyer had addressed, with how many variations of the same presentation.

After the testimony concluded, *Frank* called for the vote. He raised his own hand before fully processing that he was raising it. Four to one to deny the rezoning request. Board members cited concerns about fire department preparedness, conflict with the township's master plan, and inadequate answers to basic questions. One member called the proposal "a self-imposed environmental catastrophe."

The room erupted in applause. *Ellen* hugged the woman next to her. Someone shouted, "That's democracy!"

In the back, the developer's team was already packing up their materials, their faces carefully neutral. They didn't look surprised, *Frank* noticed. They didn't look concerned. They looked like people who had been through this before and knew what came next.

Afterward, in the parking lot, *Ellen* found him by his truck. Her hands were still shaking, the way they had been at the podium. "Thank you," she said. "I wasn't sure which way you'd vote."

"Neither was I," *Frank* admitted. "Not until my hand was up." He looked back at the community center, where the lights were going dark. "Ellen, I want to be honest with you. This isn't over. You saw their faces."

"I know." She pulled her cardigan tighter against the September night. "But at least we tried. At least we went on the record." She paused. "*Harold* wasn't here tonight."

Frank had noticed. *Harold*, who had farmed the land next to Ellen's since they were both young, who had lost his wife two years ago and whose children had long since moved away. *Harold*, whose five hundred acres lay directly in the developer's path. "I tried calling him yesterday," *Frank* said. "He didn't pick up."

"He's been avoiding everyone. I brought him casserole last week—he barely opened the door." *Ellen* looked toward the dark fields beyond the parking lot lights. "I think they've already made him an offer. A big one."

Neither of them said what they were both thinking: that the applause inside might not matter, that democracy had limits, that *Harold*'s decision was his alone to make.

Two days later, Related Digital filed a lawsuit.[1]

———

The lawsuit alleged that the township's denial was arbitrary and capricious. It sought to overturn the board's decision through the courts.

The township faced a dilemma. Defense would cost hundreds of thousands of dollars in legal fees—and the township's entire annual budget was modest by suburban standards. A prolonged fight could drain the treasury. Even if the township prevailed at trial, the developer could appeal, stretching litigation across years and mounting costs.

Related Digital had essentially unlimited resources—backed not just by Stephen Ross's billions but by the entire Stargate initiative, the five-hundred-billion-dollar AI infrastructure project announced by President Trump earlier that year.[2]

The township's lawyers advised the board that winning was uncertain. Michigan courts often defer to property owners seeking to develop their land. And even if the township won, the victory could be Pyrrhic. By the time the litigation concluded, years later, the township would be financially exhausted.

Frank spent the first week after the lawsuit unable to sleep past 4 a.m. He would lie awake in the dark bedroom, listening to his wife's steady breathing, watching the digital clock on the nightstand mark the hours in red numerals. 4:17. 4:23. 4:41. The house made its familiar sounds—the furnace cycling on, a branch scraping the gutter, the refrigerator's distant hum—and none of it was comforting anymore. The ceiling above him was the same ceiling he had stared at for thirty-two years in this house, but it looked different now. Heavier.

The township's annual budget was less than Related Digital would spend on lawyers in a month. He thought about *Ellen*, who had thanked him after the September meeting with tears in her eyes, gripping his hands like he had saved something precious. He thought about *Lily* and *Marcus*, who would inherit whatever decision he made. *Marcus* had asked him at dinner last Sunday what a data center was, and *Frank* had tried to explain, and the boy had said, "So it's like a really big computer?" Yes, *Frank* had said. It's like a really big computer that needs all the electricity in the world. The boy had nodded and gone back to his macaroni, satisfied. *Frank* had not been able to finish his own dinner.

On September 24, the board held a special meeting. The soybeans along the highway had turned yellow, ready to cut. The corn stood tall and dry, waiting. In a few weeks the combines would come through and the fields would be bare. Then winter. Then spring. Then someone else's problem.

David Landry addressed the board and the thirty residents who had come to hear the township's options.[3] Landry was a municipal attorney from Farmington Hills, a former mayor of Novi who had spent four decades handling cases like this one. The township's insurance company had sent him. He laid out the facts without sugarcoating them.

"You can fight this," Landry said. "But you need to understand what fighting means. Two to three years of litigation. Depleting the township's reserves. The possibility of losing anyway and paying their legal fees on top of yours. And the project gets built eventually regardless, just with a worse relationship and no concessions."

The options were settle, negotiate a consent judgment, or go to court. Michigan courts often deferred to property owners. The township would have to prove the project would cause severe harm to infrastructure. The studies the developer had submitted made that difficult.

"It's not what your residents wanted," Landry said. "But negotiating might be the best outcome available."

He made clear he was not telling the board how to vote. That was their decision. He was there to explain the law, not to make the choice for them.

———

Within three weeks of the lawsuit being filed, the board met again. The room was smaller this time—word had spread that the deal was done. *Frank* read a statement explaining the board's reasoning. His voice cracked once, on the word "impossible."

Then the board voted four to one to settle the lawsuit and approve the project.[4] One member who had voted to deny now wanted to keep fighting—but the math had changed, and she was outvoted. The consent agreement included concessions: well monitoring, fire department funding, noise limits, a community investment fund.[5] It was not what anyone had wanted. It was what they could get.

Local democracy spoke on September 10. Something larger answered on September 12. By October 1, the matter was settled.

Only after the settlement did the full picture emerge. In late October, Governor Gretchen Whitmer stood in Saline Township and announced "the largest single investment in Michigan history." The anchor tenant was OpenAI, the maker of ChatGPT, in partnership with Oracle.[6]

The Michigan Public Service Commission approved the power contracts in December 2025. More than 5,500 public comments were submitted, community leaders voiced overwhelming opposition, and public officials from both parties urged the Commission to slow down.[7] The Commission approved them anyway, without holding a contested hearing. Attorney General Dana Nessel criticized the process, noting that key contract details remained redacted even from her office.[7]

Saline Township is not unique. Across America, similar confrontations play out. In Augusta Township, just east of Saline, a Google data center proposal faces a referendum. A massive project called the Digital Gateway was defeated in Prince William County, Virginia, after years of litigation—twenty-four billion dollars in investment, rejected. Google abandoned a billion-dollar proposal in Indianapolis after opposition intensified. Microsoft walked away from a project in Caledonia, Wisconsin.

But these victories are exceptions. For every Digital Gateway that falls, a dozen Saline Townships acquiesce. The pattern that emerged here—rejection, lawsuit, settlement, approval—offers a template. Developers with deep pockets can outlast local governments with shallow ones.

Between 2024 and 2030, technology companies, private equity firms, and sovereign wealth funds plan to invest more than one trillion dollars in data center construction across the United States.[8] Over 600 projects. More than 131 gigawatts of planned power capacity—roughly seventeen percent of current American electricity consumption, concentrated in a few hundred facilities. Microsoft alone has committed eighty billion dollars per year.[9] Amazon, Google, and Meta each spend tens of billions more.

We are building the largest private infrastructure project in American history. Most Americans know nothing about it.

Here is the central insight of this story: the constraint on artificial intelligence is not software, not algorithms, not data, not even the specialized chips that perform the calculations. The constraint is electricity.

The companies building AI need electricity at a scale American utilities have never supplied to single customers. A gigawatt was once the output of the largest power plants; now it is the requirement for a single data center campus. That power must come from somewhere, and new generation takes years to build. Transmission lines must carry it, and upgrading those lines takes even longer—often a decade or more.

These requirements determine where data centers can be built more than any other factor—not land prices, not labor costs, not proximity to customers. Grid access drives everything. It explains why data centers cluster in Northern Virginia, where decades of infrastructure created the necessary capacity. It explains why new projects target rural areas, where transmission lines built to serve power plants can be repurposed for consumption. And it explains why a township of 2,300 people suddenly found itself at the center of a seven-billion-dollar deal.

This book traces the infrastructure that makes your ChatGPT query possible. We work backward from the token on your screen to the geopolitics that shape where data centers get built. The journey runs through inference mathematics, silicon chips, data center design, grid topology, power generation, land economics, capital flows, and political deal-making. Throughout, we return to Saline Township—not because it is the largest project or the most controversial, but because it offers a window into all the forces at play.

———

Stand in Saline Township today and you see construction equipment, security fencing, and a preserved red barn that gives the project its nickname. The barn is a sentimental touch, kept because it photographs well. It will remain while the fields around it become something else entirely.

The decisions being made now will structure possibilities for decades. The precedents being set—about power consumption, land use, environmental impact, and local control—will apply to hundreds of future projects. These trade-offs will shape what kind of communities Americans live in and what resources remain available to them.

The winners are celebrating. The losers are just beginning to understand what they lost.

Frank still lies awake some nights, the same thoughts cycling through his mind: The township's budget. The developer's lawyers. *Ellen*'s well. *Lily*'s face when he missed her birthday party for a board meeting. *Marcus* asking if the big computer was finished yet.

He drives past the construction site sometimes, on his way to nowhere in particular. The red barn stands in the middle of it all, preserved because someone decided it photographed well. He remembers when it was just a barn.

CHAPTER ONE

The Inference

A T 2:47 A.M. EASTERN STANDARD TIME on a Tuesday in January 2026, a college student in Ann Arbor sits at her desk, laptop screen glowing in the dark dorm room. She has a biochemistry exam in six hours and three chapters left to review. She types into ChatGPT: "Explain how the Krebs cycle generates ATP, like I'm cramming for an exam."

She hits Enter.

For a few dozen milliseconds, nothing happens. Her request travels as electrical signals through campus fiber, crosses into Comcast's network backbone, and arrives at a load balancer in one of Microsoft Azure's data centers. The system parses, authenticates, and routes it. Then something remarkable begins.

Fifteen miles southwest, a facility rises from Michigan farmland. Requests like hers will soon reach clusters of NVIDIA GB200 Grace Blackwell Superchips there. Today, her query routes through Azure data centers in Virginia or Texas. Within a year or two, it may flow through Saline Township instead. The infrastructure is the same; only the location changes. These chips, each worth tens of thousands of dollars, sit in dense racks with liquid cooling systems designed to move heat away as fast as the silicon creates it.

Her question has been converted into tokens—numerical representations of text, split into words and word fragments. Different models use different tokenizers, but the idea is the same: the model never sees letters, only numbers.

These numbers enter the neural network.

Inside the model, something happens that scientists still struggle to explain. The tokens pass through layers of artificial neurons—in models like GPT-3, 175 billion parameters.[10] Each token interacts with every other through a process called attention. The word "cycle" attends to "Krebs." The phrase "like I'm cramming" attends to the entire sentence, signaling: quick and practical, not thorough and academic.

The model calculates probability distributions. Given everything before, what word comes next? The answer, according to parameters numbering in the hundreds of billions, trained on terabytes of human text: "The." The model emits Token 464.

Now the model calculates again, but with "The" appended to the prompt. What comes next? "Krebs." Then "cycle." Then "is." Then "your." Then "cell's." Then "primary."

Each token requires a fresh pass through the model's parameters—hundreds of billions of values accessed and computed. Estimates vary, but a typical chatbot prompt consumes a few tenths of a watt-hour.[11,12] Her response, approximately 400 words, demands hundreds of individual predictions, each one a full pass through the neural network.

By the time she reads the first sentence, a facility like the one taking shape in Saline Township has consumed enough energy to run an LED light bulb for a couple of minutes.

This process is called inference—the atomic unit of AI work, the operation that makes ChatGPT, Claude, Gemini, and every other large language model useful. It happens billions of times per day.[13]

The student has no idea any of this is happening. Words appear on her screen, one by one, in a rhythm that mimics human typing. The explanation is clear and practical—exactly what she asked for. She highlights a passage, copies it to her notes, types a follow-up question. Another dozen tokens in, a few hundred out, another half-second of computation somewhere in America.

In the time it takes her to read the response, nearly a million other people around the world have asked ChatGPT questions of their own. A software developer in Singapore wants help debugging a Python script. A marketing manager in London needs to draft an email. A medical student in São Paulo is studying the same Krebs cycle, in Portuguese. A novelist in Tokyo brainstorms plot ideas. Each question becomes tokens. Each token triggers inference. Each inference consumes compute, power, and cooling capacity in facilities scattered across the globe.

The scale is difficult to grasp. When billions of prompts turn into hundreds of tokens each, the total volume of generated text becomes almost too large to describe.

And yet, for all this scale, the experience feels intimate. The student feels like she is talking to something, not using something. The response matches her voice, addresses her specific question, adjusts to her context of exam cramming. The infrastructure that makes this possible, from billions in capital to megawatts of power to acres of computing equipment, vanishes behind an interface as simple as a text box.

This vanishing act is both an engineering triumph and a public relations challenge. The ease of use obscures the cost. The simplicity hides the scale. And as AI becomes woven into daily life, the gap between perception and reality grows ever wider.

1.1 THE REQUEST

Her question is one of about 2.5 billion that OpenAI's systems process daily—roughly 29,000 prompts per second, before counting Copilot, Gemini, Claude, and everything else.[13] Each request triggers the same sequence: tokenization, embedding, attention, prediction, sampling. The infrastructure required to perform this operation at scale is what this book is about.

Each request demands dedicated GPU compute, consumes electricity, generates heat that must be removed. Add them together across a day, a month, a year, and the scale of investment comes into focus. Microsoft spends $80 billion annually on data center infrastructure.[9] Developers approach small farming communities across America, offering hundreds of millions of dollars for land near high-voltage transmission lines.

Inference is where the money is. Training a model like GPT-4 costs perhaps $100 million—a massive sum, but a one-time expense.[14] Operating that model for a year, handling billions of requests per month, costs far more. Microsoft Azure processed over 100 trillion tokens in the third quarter of 2025 alone, 50 trillion in a single month.[15] At current API pricing, that represents billions of dollars in compute costs, recurring every quarter, growing 5x year-over-year.

We have entered what researchers call the inference era. Training is not going away, but the financial center of gravity has shifted toward deployment. The question is no longer just "Can we train a smarter model?" It is "Can we serve that model to hundreds of millions of users simultaneously, at latencies measured in milliseconds, without the power grid collapsing?"

The answer is building itself, right now, in places like Saline Township.

To understand why inference dominates the economics of AI, we must understand how it differs from training.

Training is the process of creating a model. It involves exposing a neural network to massive amounts of text—books, websites, code repositories, scientific papers—and adjusting parameters to better predict each next word. Training GPT-4 required roughly 25,000 NVIDIA A100 GPUs running for three months, consuming perhaps 50 gigawatt-hours of electricity. Estimates put the cost at $60 million to $100 million in compute alone.[14]

But training happens once. The parameters it produces—the weights that encode what the model "knows"—are frozen and deployed. Those same weights serve every user, every request, indefinitely.

Inference, by contrast, never stops. Every time someone asks Chat-GPT a question, the frozen model springs into action. The weights are loaded into GPU memory. The input is processed through every layer. A response is generated, token by token. Then the weights wait for the next request.

Think of the difference between writing a book and reading it aloud. Writing is arduous but finite. You do it once, and the book exists. Reading that book aloud to every person who wants to hear it, individually, on demand, twenty-four hours a day—that is where the real work accumulates.

In 2022, when ChatGPT launched, training costs far exceeded inference costs. The model served millions of users, but usage was modest. By 2024, the ratio had inverted. By 2025, inference accounted for the majority of AI compute expenditure at every major hyperscaler. The models were trained; now they had to be served, at scale, to hundreds of millions of people.

This shift has profound implications for infrastructure. Training clusters can be built anywhere with cheap power and good cooling. They don't need proximity to users or low latency. They can run batch jobs overnight. Training is patient.

Inference is impatient. Users expect responses in milliseconds. Latency matters. Geographic distribution matters. Reliability matters—when the inference system goes down, users notice immediately. The infrastructure requirements for serving a model are categorically different from those for training one.

The data center buildout happening across America is primarily an inference buildout. The facilities rising in Saline Township, in Kansas

17

farmland, in Arizona desert, in Virginia suburbs—these are not training clusters. They are inference engines, designed to serve hundreds of millions of users with response times measured in hundreds of milliseconds.

1.2   WHAT HAPPENS WHEN AI THINKS

To understand why inference demands such enormous resources, we need to understand what a large language model actually does. The explanation requires simplification, not distortion.

A language model is a probability machine. Given some text, it predicts what comes next. This sounds simple. It is not.

When you type "The capital of France is," the model doesn't look up the answer in a database. It doesn't search the internet. Instead, it performs a mathematical transformation on your input, drawing on patterns learned from hundreds of billions of words of human text. The transformation produces a probability distribution over its entire vocabulary. Paris might have 94% probability. Lyon, 0.3%. Marseille, 0.2%. Nonsense tokens, tiny fractions.

The model samples from this distribution—typically selecting the highest-probability token, sometimes introducing randomness for variety—and outputs "Paris." Then it appends "Paris" to the input and repeats. "The capital of France is Paris. It" becomes the new prompt. What comes next? Perhaps "is" at 45%, "has" at 22%, "was" at 18%.

This autoregressive generation—predicting one token at a time, each prediction depending on all previous tokens—is the core of modern language AI.

The mathematics behind this process centers on a structure called the transformer, introduced in 2017 by researchers at Google in a paper titled "Attention Is All You Need." The transformer's key innovation is the attention mechanism—a way for the model to dynamically focus on relevant parts of the input when making predictions.

Consider the sentence: "The animal didn't cross the street because it was too tired." What does "it" refer to? A human knows immediately: the animal. We understand that tired applies to animals, not streets. The attention mechanism allows the model to make similar connections. When processing "it," the attention layers assign high weights to "animal" and low weights to "street." This weighted attention informs the prediction.

In practice, transformers have many attention heads, each learning different relationships. Some heads focus on grammatical structure. Others track entity references. Still others encode semantic meaning that researchers struggle to interpret. The heads work in parallel, their outputs combined and passed through additional neural network layers.

A typical large language model stacks dozens of these transformer layers. GPT-3 had 96.[10] Some frontier systems likely use around a hundred, but companies rarely disclose the details. Each layer transforms the representation of the input, building increasingly abstract features.

Early layers learn basic patterns: word boundaries, simple grammar, common phrases. Middle layers encode more complex relationships: subject-verb agreement across long distances, thematic consistency, logical structure. Upper layers capture high-level concepts: tone, style, intent, factual recall.

This layered structure creates the emergent capabilities that make modern LLMs useful. The model was never programmed to summarize text, translate languages, or write code. These abilities emerged from learning to predict the next token on massive amounts of human-generated text. The same architecture that predicts "Paris" after "The capital of France is" can generate working Python functions, explain quantum mechanics, or draft legal contracts.

But emergence comes at a cost. Each layer means more computation per token. Each token requires passing through every layer. GPT-3's 175 billion parameters must all be accessed, multiplied, and summed for every single token generated.[10]

The computational challenge of inference is not primarily about math—modern GPUs excel at matrix multiplication. The bottleneck is memory.

The model's parameters must be stored somewhere—in high-bandwidth memory (HBM) on the GPU. An H100 has 80 gigabytes of HBM.[16] A 70-billion-parameter model, stored in standard precision, requires about 140 gigabytes. More than one GPU can hold. Larger models require multiple GPUs working together, adding coordination overhead and complexity.

Even more challenging is the key-value cache. The model caches intermediate computations during token generation to avoid redundant work, and this cache grows with context length. A 70-billion-parameter model processing a 32,000-token context might need 32 gigabytes just for the cache. At 128,000 tokens—the context length of some modern models—the cache could exceed the model weights themselves.

Memory bandwidth—how fast data moves from storage to processing cores—becomes the limiting factor. An H100 can perform quadrillions of calculations per second but moves only about 3 terabytes of data per second. For inference workloads, the GPU often waits for data rather than crunching numbers. Engineers call this being "memory-bound" rather than "compute-bound."

This distinction matters for data centers. Memory-bound workloads behave differently than compute-bound ones: different heat patterns, different optimization strategies, different scaling with hardware investment. Training is often compute-bound. Inference is usually memory-bound. This shift reshapes chip design, facility architecture, and the entire economics of AI infrastructure.

What makes this remarkable is how much computation hides behind such a simple interaction. The student typed a question and received an answer. The interface is a text box. Behind that text box lies one of the most complex software systems ever built.

Consider what happens before her request even reaches the neural network. Her browser encrypts the message and sends it over HTTPS. DNS servers resolve the domain. A content delivery network routes the request to the nearest entry point. A load balancer selects a data center. An authentication service verifies her account. A rate limiter checks her usage against her subscription tier. A moderation system scans for policy violations. A routing layer selects which model to use. A scheduler assigns the request to an available GPU cluster.

Only then does inference begin.

After inference, more systems engage. Safety checks scan the response. The system logs token counts for billing, usage metrics for capacity planning. The response is compressed and transmitted. Browser-side code renders the streaming tokens.

This stack has been refined over years of engineering. Early ChatGPT deployments were fragile, slow, frequently overloaded. By 2026, the system handles billions of requests daily with remarkable reliability. That reliability comes from layer upon layer of sophisticated engineering, each adding latency, complexity, and operational overhead.

The engineering teams that build and maintain these systems number in the thousands. OpenAI employs hundreds of infrastructure engineers. Microsoft Azure's AI team is larger still. Google's DeepMind and Cloud divisions together employ thousands more. The inference stack has become one of the largest software engineering efforts in history, comparable to operating system development or large-scale database systems.

Unlike traditional software, inference systems must manage physical resources at industrial scale. They are not just code—they are power, cooling, networking, and real estate. The student's simple question triggers computation, but also heat dissipation, network traffic, and electricity consumption. The software meets the physical world in the data center.

1.3 THE ARCHITECTURE

We can now look more closely at what happens inside the transformer when the student's biochemistry question arrives. Understanding the mechanics reveals why the architecture demands such enormous resources.

———

She typed: "Explain how the Krebs cycle generates ATP, like I'm cramming for an exam."

The first step is tokenization. The model cannot process English directly; it works with numbers. A tokenizer breaks the text into subword units and assigns each a numerical ID. Her question becomes sixteen tokens. Common words like "the" get single tokens, but "cramming"—less common—splits into "cr" and "amming." The tokenizer's logic is statistical, not grammatical.

Most modern LLMs use variations of byte-pair encoding, an algorithm that dates back to 1994, originally designed for data compression. GPT-5's tokenizer recognizes about 200,000 distinct tokens—double the vocabulary of GPT-4. This vocabulary determines how efficiently text can be represented. English text typically tokenizes to about 1.3 tokens per word, though this varies with vocabulary complexity.

Each token ID maps to an embedding vector—a list of thousands of numbers representing the token's meaning in a high-dimensional space. Tokens with related meanings cluster near each other in this space. Her sixteen tokens become a matrix that enters the transformer.

———

Here is where transformers earn their name. The architecture was introduced in 2017 by researchers at Google in a paper titled "Attention Is All You Need"—one of the most consequential papers in the history of computing. Before transformers, neural networks processed sequences one element at a time, making them slow to train and prone to forgetting earlier context. The transformer's key innovation was *attention*: a mechanism that allows every token to interact with every other token simultaneously.

Consider the token "cycle" in her input. In isolation, the word is ambiguous—bicycle? Economic cycle? The attention mechanism lets the model recognize that "Krebs" is the relevant context. For "cycle," the attention layers assign high importance to "Krebs" and "generates," low importance to "exam" and "cramming." Each layer contains multiple attention heads, each learning different relationships. Some heads focus on grammatical structure. Others track entity references. By the time the input has passed through 96 or more layers, each token's representation has been enriched by context from every other token.

At the top of the network, a final transformation produces a probability distribution over all possible next tokens. "The" might have 32% probability. "ATP" might have 18%. The model samples from this distribution, emits a token, appends it to the context, and repeats.

This autoregressive generation—predicting one token at a time, each prediction depending on all previous tokens—is the core of modern language AI. It explains why responses stream in word by word rather than appearing all at once. Each word requires a full computational pass. And it explains the resource demands: the 300th token's prediction depends on all 299 before it. Caching reduces redundant computation, but the cumulative cost adds up.

The numbers compound relentlessly. A tokenizer recognizes 200,000 distinct tokens. Each maps to an embedding vector of 8,192 to 16,384 values. The model stacks 96 or more transformer layers. Multiply these together: a single forward pass through the network involves tens of trillions of arithmetic operations.

These architectural choices are not arbitrary. Larger vocabularies reduce the number of tokens needed to represent text but increase memory requirements. Deeper networks capture more complex patterns but multiply computation per token. Longer context windows enable richer conversations, but the key-value cache grows linearly with context length. Every design decision trades capability against cost.

The cumulative effect is staggering. The International Energy Agency estimates that data centers consumed 415 terawatt-hours of electricity in 2024 and could reach 945 terawatt-hours by 2030.[17] AI inference is a significant and growing fraction of this total.

Engineers wage constant war against inefficiency. Flash attention restructures memory access patterns. Continuous batching groups requests dynamically. Quantization reduces numerical precision, trading small accuracy losses for substantial speed gains. A typical large language model becomes two to five times faster to serve within a year of deployment, purely through software improvements.

But optimization has limits. Demand growth often outpaces efficiency gains. The inference infrastructure keeps expanding because the improvements, substantial as they are, cannot keep up with the explosion in usage.

1.4 THE SCALE OF THE MACHINE

Scale alone does not explain the data center buildout. Latency does.

If inference could happen anywhere, hyperscalers would consolidate facilities wherever electricity costs least—Iceland, rural Nevada, the Australian outback. They cannot. Users expect responses in milliseconds, and physics constrains where those milliseconds can be spent.

For conversational interfaces like ChatGPT, the key metric is time to first token (TTFT)—how long users wait before the response begins streaming. Ideal TTFT is under 200 milliseconds, creating the impression of instantaneous response. Under 500 milliseconds is acceptable. Beyond one second, users perceive delay; beyond three, they perceive failure.

For code completion, tolerances are tighter. Developers expect suggestions as they type, and latencies beyond 100 milliseconds disrupt flow. GitHub Copilot's infrastructure prioritizes sub-100-millisecond TTFT for single-line suggestions.

For voice assistants, latency compounds. The system must transcribe speech to text (100+ milliseconds), generate a response (200+ milliseconds), and synthesize speech from text (100+ milliseconds). Total latency must stay under one second for natural conversation.

These requirements constrain geography. A data center in Virginia cannot serve low-latency requests to users in Tokyo. The speed of light imposes a floor of about 80 milliseconds for transpacific round trips. Serving global users with consistent latency requires geographically distributed infrastructure—including facilities in places like Saline Township.

Latency and cost exist in tension. Lower latency requires dedicated GPU capacity held in reserve; cost efficiency requires maximizing GPU utilization through batching—processing many requests together. Production systems navigate this tradeoff dynamically: premium tiers get dedicated capacity, free tiers get batched requests.

The economics are ruthless. If a provider reduces cost per token by 10%, that translates to hundreds of millions in annual savings. If latency improvements increase user engagement by 5%, that means hundreds of millions in additional revenue. Tokens are not just a technical artifact—they are the unit of value creation in the AI economy.

Every token consumes electricity. Every joule of electricity becomes heat. Every watt of heat must be removed by cooling systems that themselves consume electricity.

An H100 GPU draws about 700 watts under full load.[16] A facility serving significant AI traffic might deploy 10,000 GPUs—7 megawatts for GPUs alone, 15 to 20 megawatts total with cooling overhead. Microsoft's total AI infrastructure operates at capacity measured in gigawatts.

Unlike many industrial processes, AI inference cannot be scheduled for off-peak hours. The student studies at 2:47 a.m. and expects an answer immediately. The software developer in Singapore is in the middle of her workday. AI data centers run at constant load around the clock. Users in North America sleep while users in Asia and Europe work. For utilities, this constant demand means no opportunity for scheduled maintenance, no off-peak periods for grid stress relief. The always-on nature of inference shapes everything about how these facilities integrate with the power grid.

1.5 Who Sees What in the Inference

The same token generation process looks different depending on where you stand—and these differences create tensions that ripple through the rest of this book.

For machine learning engineers, the token is a benchmark. Tokens per second per dollar. Throughput under latency constraints. They optimize for efficiency because every percentage point of improvement translates to millions saved at scale. Their incentives align with technical performance, not community impact.

For technology executives, the token is a unit of revenue. Pricing, costs, and margins are calculated per token. When Microsoft spends $80 billion annually on infrastructure, the calculation behind that spending is inference capacity—how many tokens can we serve, at what cost, at what margin? Their incentives align with growth and return on capital.

For the student in Ann Arbor, the token is invisible. She neither knows nor cares that her question triggered computation on thousands of GPU cores. She cares that the response helps her exam. Good infrastructure is invisible—the complexity hidden, only the utility remaining. Her incentives align with convenience and cost.

But users are also citizens, voters, and ratepayers. The infrastructure required to serve them is being built in communities they may never visit, powered by utilities that serve their neighbors, enabled by incentives funded by taxpayers. The disconnect between user and infrastructure creates a political problem: those who benefit most from AI often live far from those who bear its costs.

For residents of Saline Township, inference is not an abstraction but a transformation—the physical, social, and economic remaking of their community.[18] Some see opportunity: construction jobs, permanent positions, increased tax revenue. Others see disruption: farmers facing offers they cannot refuse, neighbors worried about noise and traffic, environmental advocates questioning water consumption.

These perspectives do not naturally align. Engineers optimize for throughput. Executives optimize for margin. Users optimize for convenience. Communities absorb externalities. The political economy of AI infrastructure—the subject of later chapters—emerges from these misaligned incentives. The student in Ann Arbor and the farmer in Saline Township are connected by tokens flowing through fiber optic cables, but their interests diverge in ways that zoning boards, utility commissions, and state legislatures must somehow reconcile.

1.6 INFERENCE EVERYWHERE

Four forces are driving AI inference demand faster than any computing workload in history.

First, model capabilities keep improving, expanding the range of useful applications. Tasks that AI could not perform well two years ago—complex coding, nuanced writing, multimodal understanding—are now viable for mainstream deployment.

Second, integration into existing products multiplies usage. When GitHub Copilot becomes a default feature, inference requests per developer jump from zero to hundreds per day. When Outlook adds AI summarization, every email recipient becomes an inference customer. When Windows integrates Copilot, every PC user generates requests.

Third, new AI-native applications emerge. Insurance claims processing, financial analysis, marketing copywriting—entire industries are rebuilding workflows around AI inference. Each workflow replacement creates sustained, high-volume inference demand.

This is the economic function of inference, stated plainly: converting electricity into cognitive output. Manufacturing automation converted electricity into physical labor—assembly, fabrication, movement. AI inference converts electricity into something that resembles thinking: analysis, communication, judgment, recommendation. The kilowatt-hours flowing into Saline Township produce not steel or automobiles but the functional equivalent of work that previously required human minds.

The implications extend beyond the 450 permanent jobs the facility will create. A Stanford study found that entry-level workers in AI-exposed occupations experienced a 13 percent relative decline in employment between late 2022 and mid-2025.[19] Software developers aged 22 to 25 saw employment fall nearly 20 percent from its peak. Workers over 30 fared better; their experience remained valuable even as AI handled routine tasks. But the pattern was clear. The same infrastructure that creates a few hundred technician jobs in rural Michigan enables the displacement of office workers around the world.

This is not the displacement pattern of previous technological revolutions. Manufacturing automation primarily affected blue-collar workers without college degrees. AI disproportionately affects the educated, the urban, the white-collar—the demographic most likely to use the very tools that threaten their employment.[20] The paralegal reviewing contracts. The customer service representative answering calls. The junior analyst building spreadsheets. The medical coder assigning billing codes. These are the occupations where AI capability already meets or exceeds human performance on routine tasks. The infrastructure rising in places like Saline Township is not just serving students cramming for exams. It is producing cognitive labor at industrial scale.

Fourth, agentic AI architectures multiply requests. An AI agent that plans, executes, and iterates might generate thousands of tokens internally for each user-visible response. Early experiments suggest agents could consume 10 to 100 times more inference than simple question-answering.

Projections vary, but most analysts expect AI inference demand to grow 5 to 10 times by 2028. If efficiency improves 2x while demand grows 5x, net capacity must still grow 2.5x. That means 2.5 times as many data centers, 2.5 times as much power, 2.5 times as many GPUs. The numbers reach into hundreds of gigawatts and hundreds of billions of dollars.

Some analysts believe this growth is unsustainable—power grid constraints, semiconductor production limits, the sheer scarcity of suitable sites. Others believe growth will continue because the applications are too valuable to forgo. The infrastructure buildout underway represents one of the largest industrial investment waves in history, and later chapters examine whether the constraints will bend or break.

Every token has a carbon footprint. Every inference operation consumes electricity, and that electricity has to come from somewhere.

OpenAI, Microsoft, Google, and other major AI companies have made ambitious sustainability commitments. Microsoft pledged to be carbon negative by 2030.[21] Google committed to 24/7 carbon-free energy.[22] The reality is more complicated. AI inference demand grows faster than renewable energy supply. The marginal electricity powering a new data center in 2026 typically comes from natural gas plants, not solar farms. A commitment to purchase renewable energy certificates does not change the physics of what power plant ramped up to serve the load.

Water consumption raises similar concerns. AI data centers run hotter and denser than traditional facilities, often requiring more cooling. In water-stressed regions like the American Southwest, data center water use competes with agriculture, municipal supply, and environmental needs. Chapter 6 examines the generation mix in detail; Chapter 7 explores how these environmental concerns intersect with land use decisions.

For the student asking about the Krebs cycle, these costs are invisible. Her question arrives, an answer appears. The carbon emitted, the water consumed, the land transformed—these are externalities, borne by someone else, somewhere else. The tokens are not free.

1.7 FIFTEEN MILES SOUTHWEST

Return to Saline Township, where the technical abstractions of this chapter take physical form.

The site was chosen for reasons we explore in later chapters: available transmission capacity, cooperative regulatory environment, proximity to skilled labor, and land that could be assembled quickly without brownfield complications.

Inside the facility, a hierarchy of scale will organize the computing equipment. The GB200 Superchips sit in rack-scale systems, seventy-two processors per cabinet. The cabinets are arranged in rows. The rows are organized into pods. The pods form data halls. The data halls fill buildings. The buildings cover the campus. The campus will consume 1.4 gigawatts.

This hierarchy matters because failure cascades. A dead GPU means a sled runs at reduced capacity. A dead sled means a rack loses capability. A dead rack means requests route elsewhere. A dead data hall—power failure, cooling failure, network failure—means thousands of users experience degraded service.

Redundancy is engineered at every level. Dual power feeds from independent substations. Backup generators with days of fuel. Multiple cooling systems with diverse water sources. Network connections through independent fiber paths. The goal is what the industry calls "five nines" availability: 99.999 percent uptime, or about five minutes of downtime per year.

The term "five nines" comes from the telephone industry, where AT&T set the standard for service reliability in the 20th century. If your phone worked 99.999% of the time, you never thought about it; it was simply there. The telecom engineers who built those systems, some of whom later moved into data center design, brought that expectation with them. When the student types her question at 2:47 a.m., she assumes an answer will appear. That assumption rests on decades of reliability engineering, now applied to facilities consuming more power than small cities.

Achieving five nines at gigawatt scale is an engineering challenge without historical precedent. The Saline Township facility represents the frontier of industrial infrastructure—the complexity of a large power plant, the precision of a semiconductor fab, the scale of a major logistics center. Chapter 3 examines the engineering in detail.

What matters beyond the engineering is the human geography.

The 450 permanent employees who will staff the facility include electrical engineers, mechanical engineers for cooling systems, network engineers, security personnel, and administrative staff. Few will be AI experts—the machine learning engineers who design the models work in San Francisco, not Michigan. High-paying AI jobs cluster in expensive cities. Operational jobs locate where data centers are built.

The construction workforce is larger but temporary. At peak, 2,500 workers pour concrete, install electrical systems, and mount racks. Many are specialists who move from project to project across the country. When construction ends, they move on.

This labor geography reflects a broader pattern. The student in Ann Arbor who types a question benefits from AI. The engineers in San Francisco who design the models benefit from AI. The investors in New York who finance the buildout benefit from AI. The community in Saline Township hosts the infrastructure, absorbs the disruption, and debates whether 450 permanent jobs offset the transformation of their farmland.

Whether this exchange is fair—and who gets to decide—is a question that zoning boards, utility commissions, and state legislatures are only beginning to confront. The technical architecture of inference creates winners and losers. The political economy determines who ends up where.

1.8 THE HARDWARE QUESTION

Inference consumes enormous resources, but those resources take specific forms. Her token was generated by NVIDIA GPUs—Blackwell-generation chips costing $60,000 to $70,000 each, drawing over a kilowatt of power, representing the leading edge of semiconductor engineering.

These chips are not fungible. You cannot generate ChatGPT responses on ordinary processors. The transformer architecture demands specific capabilities: massive parallel processing, high memory bandwidth, specialized tensor operations. Only a handful of chip designs in the world meet these requirements. NVIDIA dominates with roughly 80 percent market share.[23]

This dominance creates bottlenecks. Every AI company wants Blackwell chips. Production is constrained by advanced semiconductor manufacturing capacity, concentrated in Taiwan. Export controls limit which countries can purchase the most advanced chips. Geopolitical tensions threaten supply chains.

The chips also drive data center design. Their power demands dictate cooling requirements. Their form factors determine rack configurations. Their failure rates shape redundancy architectures. The building in Saline Township is, in a very real sense, designed around the chips it houses.

To understand the data center buildout, we must understand the silicon that fills those data centers. The next chapter examines the hardware revolution that makes AI inference possible—and the bottlenecks that make it so challenging to scale.

———

Every technology revolution has required infrastructure. The automobile needed roads and gas stations. The telephone needed wires and switches. The internet needed routers and data centers. The AI buildout is happening faster than any of these predecessors—attempting similar scale in less than a decade.

The student in Ann Arbor sits at the end of a vast supply chain. Her question triggers computation spanning the globe: silicon fabricated in Taiwan, assembled in Texas, powered by natural gas from Pennsylvania, cooled by Michigan aquifers, financed by Wall Street, enabled by policies shaped in Lansing and Washington. She knows none of this. The magic of modern infrastructure is that it works invisibly.

This book is an attempt to make the invisible visible—to trace the infrastructure that makes AI possible, from the token to the power plant, from the GPU to the transmission line, from the data center to the zoning board. Some researchers worry about more than infrastructure. They ask what happens when systems trained in these facilities become capable enough to design their own successors—a question this book does not attempt to answer, but one that shadows every gigawatt deployed. Inference is the starting point. Everything else follows from the need to perform this operation, at scale, with speed, reliability, and efficiency.

The token has been generated. Now let us understand how.

CHAPTER TWO

The Silicon

THE CLEAN ROOM AT TSMC's FAB 18 operates in perpetual twilight, yellow-orange lights filtering out the wavelengths that might damage photosensitive wafers. At 3:17 AM Taiwan time, a process engineer named *Wei-Lin*—a composite representing the technicians who build the world's most advanced semiconductors—watches a silicon wafer slide into the extreme ultraviolet lithography machine. The wafer contains dozens of NVIDIA B200 dies, each one destined for an AI data center somewhere in America.

Wei-Lin has worked this shift for six years. He knows the EUV machine's rhythms the way a pilot knows an aircraft. The tool costs $200 million and takes two years to deliver. Inside, a laser pulse vaporizes molten tin droplets fifty thousand times per second, generating ultraviolet light at a wavelength of 13.5 nanometers—roughly the width of a few dozen atoms. That light reflects off mirrors polished to atomic smoothness and etches patterns onto silicon at resolutions that seemed impossible when *Wei-Lin* started his career.

The wafer emerging on the other side carries transistors measured in nanometers. 208 billion of them on each B200 die, packed into an area smaller than a playing card.[24] In industry parlance, these finished chips are simply called "silicon"—a term that has come to mean the manufactured product, not the raw element from which it starts. These

35

transistors—electronic switches that form the building blocks of all digital logic—will eventually perform the mathematical operations that make artificial intelligence possible. But first they must survive a journey of seventy more processing steps, each one capable of destroying the entire batch.

"One particle," *Wei-Lin* says, watching the wafer's progress on his monitor. "One contamination event. Millions of dollars gone." He has seen it happen. A speck of dust smaller than a human cell, landing in the wrong place, can ruin chips that would have sold for fifty thousand dollars each.

The chips being manufactured on this shift will eventually reach Saline Township, Michigan, where they will be installed in racks consuming 140 kilowatts each—enough electricity to power over a hundred American homes at any given moment.[25] But between the clean room in Taiwan and the data center in Michigan lies a supply chain that spans three continents, involves dozens of specialized manufacturers, and represents the most complex industrial process ever created. This chapter traces that journey.

2.1 ARCHITECTURE OF INTELLIGENCE

To understand why a single chip costs $50,000 and why companies are spending billions to acquire them, we need to understand what these chips actually do. The answer lies in a particular kind of mathematics: matrix multiplication.

When an AI system like ChatGPT processes your question, it performs billions of mathematical operations. Not complex calculus or abstract algebra—just multiplication and addition, repeated at incomprehensible scale. A typical interaction with a large language model involves multiplying matrices containing hundreds of billions of numerical values. The speed at which a computer can perform these operations determines how fast the AI can think.

Traditional computer processors—the CPUs that power laptops and smartphones—are designed for versatility. They run spreadsheets, browse the web, edit videos, execute millions of different programs. This flexibility comes at a cost: they process information sequentially, one instruction at a time. A modern Intel or AMD processor might have sixteen or thirty-two cores, each capable of independent work, but the fundamental design prioritizes breadth over depth.

Graphics Processing Units evolved differently. Originally designed to render video games, GPUs contain thousands of smaller, simpler processing cores that can work simultaneously. A video game needs to calculate the color and brightness of millions of pixels on screen sixty times per second. Each pixel calculation is independent of the others, making them ideal for parallel processing. An NVIDIA GPU might contain over 16,000 cores, each less powerful than a CPU core individually but collectively capable of performing far more calculations per second when the work can be divided among them.

In 2012, researchers at the University of Toronto demonstrated that GPUs could dramatically accelerate a type of AI called deep learning. Alex Krizhevsky, Ilya Sutskever, and Geoffrey Hinton used two NVIDIA GeForce GTX 580 graphics cards—consumer hardware designed for video games—to train an image recognition system that outperformed every previous approach.[26] The advantage was not incremental. Their system, AlexNet, cut the error rate for image classification nearly in half.

That breakthrough launched an arms race. The same mathematical operations that rendered video game graphics—matrix multiplications performed in parallel—turned out to be exactly what neural networks needed. NVIDIA, which had been primarily a gaming company, suddenly found itself at the center of the AI revolution.

Jensen Huang, NVIDIA's CEO since its founding in 1993, recognized the opportunity earlier than his competitors. In 2006, the company released CUDA, a programming framework that allowed developers to use

GPUs for general-purpose computing beyond graphics. When the deep learning revolution arrived, NVIDIA had a six-year head start in software tools and developer relationships. That lead proved decisive.

I was at the University of Michigan when CUDA arrived, working at the Center for the Study of Complex Systems. We installed some of the first CUDA-capable cards the university deployed, using them for agent-based modeling and particle simulations. The idea that these chips would, within fifteen years, power systems capable of passing the bar exam seemed absurd. It was not. It was simply early.

The scale of improvement since 2012 defies ordinary intuition. AlexNet trained on two GPUs with 3 gigabytes of memory each. Thirteen years later, frontier AI systems training on clusters containing hundreds of thousands of GPUs, each with memory measured in hundreds of gigabytes. Total computing power for training has increased by a factor exceeding one hundred million. This is not gradual progress. It is explosive growth unlike anything in the history of technology.

This growth did not happen automatically. It required designing chips specifically for AI workloads, building software platforms that made those chips usable, and constructing data centers capable of housing and cooling them. Each element of the stack evolved together, creating a technological system that consumes resources on industrial scales while producing capabilities that seemed impossible a decade ago.

2.2 How NVIDIA Conquered AI

By 2025, NVIDIA controlled roughly 80 to 90 percent of the market for AI accelerators, depending on how the market is defined.[23] No other technology company enjoyed such complete dominance in a sector this important. Understanding how this happened requires examining three interlocking advantages: technology, software, and supply chain control.

NVIDIA's flagship chips—the Blackwell family, which succeeded the Hopper generation (H100 and H200)—contain specialized circuits called Tensor Cores, designed exclusively for the matrix operations that AI requires. These are not general-purpose processors that happen to work well for AI. They are purpose-built accelerators, optimized at the silicon level for the specific mathematical patterns of neural networks.

The progression from 2020 to 2025 follows a consistent pattern: each generation roughly triples AI performance while doubling power consumption and price. What matters is what this buys. The A100 in 2020 could hold a 7-billion-parameter model in memory. Parameters are the individual learned settings that shape how the model responds, like billions of tiny dials tuned during training. By 2025, the B300 holds models approaching 100 billion parameters on a single chip. Context windows—how much text a model can consider at once—expanded from thousands of tokens to hundreds of thousands, because the key-value cache that grows with context finally had room to breathe. Training runs that required thousands of chips coordinated across a datacenter can now fit on hundreds, reducing the networking complexity that dominates large-scale AI systems.

The economics shifted with each generation. When inference meant running the A100 at 400 watts, operators optimized for chip count—pack more GPUs into fewer racks. The B300's 1,400-watt appetite inverts this logic. Power and cooling now constrain density more than physical space. A rack that once held eight A100s might hold four B300s, but those four deliver more than ten times the throughput. The constraint moved from "how many chips can we buy" to "how many chips can we power." This is why datacenter design changed so dramatically between 2020 and 2025: the silicon demanded it.[1624]

Raw capacity is only half the story. Memory bandwidth—how fast data moves between memory and processor—often matters more. The B300 transfers data at over five terabytes per second, fast enough to stream a full-length movie in under a millisecond. This bandwidth comes from stacking memory vertically on the same package as the GPU, connected through thousands of tiny pathways called through-silicon vias. The architecture reduces the distance data must travel from centimeters to micrometers. Only a handful of companies can manufacture these packages—all of them in Taiwan and South Korea.

Blackwell introduced another innovation: connecting two GPU dies on a single package through NVIDIA's NVLink-C2C interconnect. Previous generations used a single monolithic chip. Blackwell uses two chips working in concert, appearing to software as a unified processor. This approach circumvents manufacturing limitations on chip size while enabling more transistors per package. The B200 contains 208 billion transistors—more than any chip previously manufactured—and the packaging technology opens possibilities for even larger systems in future generations.

This relentless improvement creates both opportunity and anxiety. Organizations that deploy current hardware gain competitive advantages, but only temporarily. A cluster purchased in 2023 was obsolete by 2025— not broken, just outclassed. The companies that trained GPT-4 on A100s watched competitors train better models on H100s in less time for less money. The next generation is always eighteen months away, and waiting means falling behind. This dynamic explains why AI companies sign multi-billion-dollar contracts for chips that do not yet exist: the alternative is irrelevance.

Hardware alone does not explain NVIDIA's dominance. The company's software platform creates switching costs that render competitors' hardware advantages irrelevant for most customers.

CUDA—Compute Unified Device Architecture—is NVIDIA's proprietary programming framework. Nearly every major AI framework, from

PyTorch to TensorFlow, runs best on CUDA. Researchers and engineers have spent years mastering CUDA code. Libraries of pre-trained models, debugging tools, and optimization techniques all assume NVIDIA hardware.

This creates a chicken-and-egg problem for competitors. Even when rival chips offer compelling specifications at lower prices, developers must rewrite their code for alternative platforms that lack the polish, documentation, and community support CUDA has accumulated over two decades. For most organizations, retraining engineers and rebuilding software pipelines costs more than any hardware savings.

Google's Tensor Processing Units face the same challenge. Custom-designed for AI workloads and available through Google Cloud, TPUs offer excellent performance for models built on TensorFlow. But the broader AI community has standardized on PyTorch, which runs best on NVIDIA hardware. Google uses TPUs extensively for internal development; external adoption remains limited.

NVIDIA does not manufacture its own chips. The company designs architectures and outsources production to Taiwan Semiconductor Manufacturing Company (TSMC), the world's most advanced chip foundry. This "fabless" model—designing chips without owning factories—allows NVIDIA to focus on architecture while TSMC invests tens of billions in manufacturing capacity. AMD, Apple, and Qualcomm operate the same way. Intel took the opposite approach for decades, designing and manufacturing its own chips, but has struggled to match TSMC's manufacturing precision and now outsources some production as well.

But NVIDIA controls something equally valuable: allocation. When demand exceeded supply—as it did dramatically from 2023 through 2025—NVIDIA decided who received chips and how many. The company established tiers, prioritizing strategic partners: cloud providers like Microsoft

Azure and CoreWeave, AI companies like OpenAI and xAI, and government research institutions.

Smaller companies and startups found themselves at the back of the queue, waiting six to twelve months for chips while competitors with better NVIDIA relationships scaled their AI capabilities. Some turned to secondary markets, paying premiums of 60 to 100 percent above list price—when chips were available at all.

This allocation power gives NVIDIA influence beyond market share. Partners who maintain good relationships receive priority access to next-generation chips. Those who complain publicly, or invest heavily in competing platforms, risk falling down the list. The cycle is self-reinforcing: dominance ensures continued dominance.

The financial results demonstrate this dominance. NVIDIA's data center revenue grew from $10.6 billion in fiscal 2023 to $47.5 billion in fiscal 2024—growth of 348 percent in a single year.[27] The company's total revenue reached $60.9 billion, with margins that would be remarkable for software companies and are extraordinary for hardware.[27] Gross margins on data center products exceed 70 percent, reflecting both strong demand and limited competition.[27]

This revenue concentration creates dependencies across the AI industry. The largest AI companies—OpenAI, Anthropic, Google, Meta—collectively purchase tens of billions of dollars in NVIDIA hardware annually. Cloud providers build their AI offerings on NVIDIA GPUs. A supply disruption would immediately constrain AI development across the entire industry. No other technology company occupies such a central position in a sector this strategically important.

2.3 THE CHALLENGERS

What would it take to break NVIDIA's hold? The challengers offer a natural experiment: better specs, lower prices, different architectures—and still they struggle.

AMD's MI300X offers 192 gigabytes of memory versus the H100's 80, at prices 20 to 30 percent lower.[28] Meta runs Llama 405B entirely on MI300X chips. Crusoe Energy placed a $400 million order.[29] Yet AMD's market share remains in the low single digits. Superior hardware is necessary but not sufficient.

Google and Amazon built custom chips for their own platforms. Google's TPUs span seven generations, with TPU v6 Trillium claiming 4.7 times the compute of its predecessor.[30] Amazon's Trainium promises 50 percent lower training costs than NVIDIA instances.[31] Both require proprietary software stacks that add friction for developers accustomed to CUDA. Both are available only through their respective clouds.

Chinese companies, blocked from buying cutting-edge chips by export controls, are accelerating domestic alternatives. Huawei's Ascend 910B provides sufficient capability for many training tasks. Alibaba and Baidu deploy Ascend infrastructure at scale. Whether Chinese chips can reach parity depends on whether China can replicate the manufacturing equipment it cannot buy—ASML's extreme ultraviolet lithography machines remain subject to export controls. Chapter 10 examines the strategic implications.

Inference specialists present a different kind of challenge. Groq—the chip company, not to be confused with xAI's chatbot Grok—builds Language Processing Units that deliver hundreds of tokens per second on Llama models, using on-chip SRAM instead of external memory.[32] NVIDIA's twenty billion dollar licensing deal with Groq in December 2025, which brought Groq's CEO and key engineers to NVIDIA while leaving the company nominally independent, suggests the threat was taken seriously.[33] Intel's Gaudi 3 offers roughly half the price of an H100 with competitive inference performance.[34]

Yet Intel's projected 500 million dollars in 2025 Gaudi sales barely registers against NVIDIA's 51 billion dollar quarterly data center revenue.[35] The CUDA software stack remains formidable: four million developers,

40,000 companies, nearly two decades of optimized libraries. For now, NVIDIA's position holds. But as inference grows to represent two-thirds of AI compute demand, facilities optimized for 140-kilowatt training racks may prove overbuilt for the chips that actually run production AI.[36]

2.4 THE EFFICIENCY QUESTION

The trillion-dollar infrastructure buildout assumes AI will continue requiring exponentially more compute. But what if it does not? Dramatically more efficient training methods and smaller, faster models raise questions about whether current projections will prove correct.

On January 20, 2025, a Chinese AI lab called DeepSeek released a reasoning model that matched or exceeded OpenAI's best on key benchmarks.[37] The performance was remarkable. The reported training cost was more so.

DeepSeek claimed to have trained its model for roughly $5.5 million.[38] A fraction of the hundreds of millions typically spent on frontier models. The claim warrants scrutiny—the figure excludes prior research, failed experiments, and hardware already owned—but even accounting for omissions, DeepSeek achieved frontier performance with dramatically less compute than American labs. On January 27, 2025, NVIDIA lost $589 billion in market cap, the largest single-day loss in U.S. stock market history.[39]

DeepSeek's efficiency came from aggressive application of known techniques: architectural innovations that activate only a fraction of the model's parameters for each query, numerical methods that reduce precision without sacrificing accuracy, clever memory management. None of these were secret. What DeepSeek demonstrated is that chip constraints had forced Chinese labs to innovate where American labs threw money.

"Money has never been the problem for us," DeepSeek's founder Liang Wenfeng told reporters. "Bans on shipments of advanced chips are the problem."[40] Export controls intended to slow Chinese AI development may have accelerated algorithmic efficiency instead. As these techniques diffuse to more players—abroad and domestically—and as innovation continues to decrease the costs of both training and inference, some of the demand assumptions underlying the trillion-dollar buildout may prove over-optimistic.

If AI becomes more efficient, will infrastructure requirements shrink? History suggests not.

The Jevons paradox, identified in 1865, observes that improvements in resource efficiency often increase total consumption. When coal-burning steam engines became more efficient, coal consumption rose as new uses became economical. Fuel-efficient cars meant people drove more miles. Cheaper LED bulbs meant buildings installed more lights.

A 2025 study in Nature Cities found the pattern holds for AI: "algorithmic efficiency gains in metropolitan data centers may enlarge, and not shrink, the energy footprint of artificial intelligence."[41] Lower costs enable new applications, larger models, wider adoption. Demand grows faster than efficiency gains.

Microsoft CEO Satya Nadella invoked this directly after DeepSeek's announcement: "Jevons paradox strikes again! As AI gets more efficient and accessible, we will see its use skyrocket."[42]

McKinsey forecasts AI-ready data center demand growing 33 percent annually through 2030.[43] The trillion-dollar buildout proceeds on the assumption that demand growth will overwhelm efficiency gains.

But the Jevons paradox is not a law. It depends on how much latent demand exists at lower price points. If AI applications are already saturated—if everyone who wants AI already has access—then efficiency gains could reduce total consumption. The question is empirical, not theoretical. The answer will determine whether the infrastructure being built today proves essential or excessive.

———

Even if total compute demand keeps growing, efficiency improvements affect where that compute happens. Techniques that reduce model size—quantization, distillation, mixture-of-experts—make inference feasible on smaller hardware. A model that once required a rack of H100s might run on a single chip after aggressive optimization.

This shifts workloads toward the edge. Apple's on-device models handle most requests without touching data centers. Qualcomm's Snapdragon X2 Elite runs capable models entirely on laptop hardware. Enterprise deployments using vLLM and Ollama avoid cloud costs by running inference locally.

The implications for centralized data centers remain uncertain but consequential. Training will likely stay centralized—the largest models still require thousands of coordinated GPUs. But inference, projected to reach 30 to 40 percent of total data center demand by 2030, could increasingly distribute to edge locations and personal devices.[44]

Facilities designed for training have different requirements than those built for inference. The liquid-cooled, 140-kilowatt racks optimized for GPU training may be overbuilt for inference workloads that run efficiently on lower-power specialized chips. Investors who bet exclusively on centralized GPU infrastructure may find their assumptions challenged by a more distributed future.

The efficiency question has no definitive answer. But it introduces uncertainty into projections that assume only growth. Efficiency improvements are real and accelerating. Their effects on total compute demand, and on its geographic distribution, remain to be seen.

2.5 THE RACE FOR CHIPS

Global demand for AI chips has created supply dynamics unlike anything the technology industry has seen. Every major technology company, dozens of AI startups, cloud providers, government laboratories, and sovereign nations compete for the same limited production capacity. Three constraints—manufacturing concentration, geopolitics, and talent—shape who gets chips and when.

NVIDIA's allocation hierarchy, described earlier, determines access to new chips. But even those with strong relationships face supply limits. During the peak of the H100 shortage in 2023, chips that listed for $25,000 to $30,000 sold on secondary markets for $40,000 to $60,000. Buyers received no warranty and faced risks of counterfeit or refurbished units. By late 2025, secondary premiums collapsed for H100 and H200 as production caught up. But Blackwell chips remain scarce, and the cycle repeats.

Every advanced AI chip depends on Taiwan Semiconductor Manufacturing Company. TSMC produces silicon for NVIDIA, AMD, Apple, Qualcomm, and dozens of other designers. No other foundry matches TSMC's capabilities at the leading edge.

This concentration creates vulnerability. Taiwan sits 100 miles from mainland China, across a strait Beijing considers Chinese territory. Any disruption to TSMC—natural disaster, military conflict, geopolitical pressure—would halt production of the most advanced chips in the world. There is no backup. Samsung operates advanced fabs in South Korea, but its technology trails TSMC's. Intel is attempting to re-enter advanced manufacturing but remains years behind, its newest process still struggling with yields.[45]

The United States has invested over $50 billion through the CHIPS Act to build domestic semiconductor manufacturing.[46] TSMC is constructing fabs in Arizona. Samsung is expanding in Texas. Intel is building in Ohio. But these facilities will not reach full production until the late 2020s, and even then will produce only a fraction of global capacity.

For now, the AI chip supply chain runs through Taiwan. Every H100, every B300, every AMD MI300X starts as silicon wafers in Taiwanese fabs. The geopolitical implications of this dependence shaped the export control regime that defines the second major constraint on global AI infrastructure.

In October 2022, the Biden administration imposed sweeping restrictions on exports of advanced AI chips to China.[47] The Federal Reserve estimates the United States maintains approximately 74 percent of global AI computing capacity, compared to China's 14 percent.[48] Whether this gap persists depends on factors examined in Chapter 10: export enforcement effectiveness, the pace of Chinese domestic chip development, and the policy decisions of successive administrations.

Demand for engineers with AI chip expertise dramatically exceeds supply. NVIDIA, AMD, Google, Apple, and startups compete for the same small pool of talent. Universities produce perhaps a few thousand graduates annually with the skills needed for GPU architecture design. This scarcity drives salaries into the hundreds of thousands for mid-career engineers, with total compensation exceeding a million dollars annually for senior architects.

The concentration of expertise creates fragility. A handful of people at NVIDIA, AMD, and TSMC possess institutional knowledge that cannot be easily replicated. Their decisions shape the entire AI industry. This concentration—in a small number of people, companies, and countries—represents a systemic vulnerability that no amount of capital can quickly address.

2.6 THE DENSITY REVOLUTION

More powerful chips requiring more cooling have transformed data center design from a real estate business into a thermal engineering challenge.

The trajectory is stark. From 2000 to 2015, web server racks ran at five to ten kilowatts, cooled by simple air circulation. Cloud computing from 2015 to 2022 pushed densities to fifteen or twenty kilowatts with optimized airflow and hot-aisle containment. Early AI deployments using A100 GPUs required hybrid approaches at thirty to fifty kilowatts. Current AI training infrastructure using H100 and B200 chips runs at 100 to 140 kilowatts per rack, demanding full liquid cooling. The next generation—projections for 2026 and beyond—anticipates 200 to 300 kilowatts per rack, requiring immersion cooling where servers sit submerged in dielectric fluid.

When Meta built its AI Research SuperCluster in 2022—16,000 NVIDIA A100 GPUs—individual racks ran at twenty to thirty kilowatts, the edge of what air cooling could support.[49] Two years later, Meta's 24,576-GPU clusters ran at over a hundred kilowatts per rack. Air cooling was gone.

NVIDIA's GB300 NVL72 system exemplifies where this trend leads. A single liquid-cooled rack containing 72 Blackwell Ultra GPUs, 36 Grace ARM processors, and 18 networking units.[50] It draws 140 kilowatts, produces 1.1 exaflops of computing performance, and cannot be deployed in any traditional data center.[50] Customers cannot order partial racks or mix configurations. For enterprise AI customers, NVIDIA no longer sells standalone chips—the product is a rack-scale system that requires a purpose-built facility.

The transition has stranded billions of dollars in legacy data centers designed for 15-kilowatt air-cooled racks. Floor loading requirements change. Ceiling heights must increase. Electrical systems need complete redesign. Operators face a choice: heavy investment in retrofits, or watching customers leave for purpose-built competitors. Coolant Distribution Units have become as scarce as the chips they cool, with manufacturing lead times stretching to six months or more. A facility with power and space but no cooling systems is useless.

CyrusOne now offers facilities capable of 300 kilowatts per rack, combining direct liquid cooling with immersion technologies.[51] Chapter 3 examines these systems and the facility architectures they require.

High-density computing also creates networking challenges. Training a large AI model requires thousands of GPUs exchanging data continuously, with bandwidth measured in hundreds of terabits per second. NVIDIA's NVLink connects GPUs within a rack at 1.8 terabytes per second. Between racks, xAI's Colossus deployment proved that hundred-thousand-GPU clusters can run on Ethernet rather than proprietary InfiniBand, potentially breaking NVIDIA's lock on networking. The networking bill for a major deployment can reach hundreds of millions of dollars. Poor network design can waste computing capacity worth billions.

2.7 THE ECONOMICS OF AI HARDWARE

The cost of AI hardware shapes who can participate in the AI revolution and who cannot. Understanding these economics requires examining capital and operating costs across the entire lifecycle of GPU infrastructure.

A single NVIDIA B300 GPU costs fifty to seventy thousand dollars. But the chip alone is useless. It must be integrated into a server with supporting processors, memory, storage, and networking. That server costs $350,000 to $600,000 for an eight-GPU configuration—thirty-seven to seventy-five thousand dollars per GPU in integration costs alone. Beyond the server, each GPU requires its share of supporting infrastructure: ten to fifteen thousand in networking equipment, five to ten thousand for liquid cooling, five to eight thousand in power distribution, two to four thousand in facility overhead. Add it up and the total deployed cost per B300 ranges from 109 to 182 thousand dollars—two to three times the sticker price of the chip itself.

These costs scale dramatically. A hundred-thousand-GPU deployment—comparable to xAI's Colossus or the first phase of Meta's Prometheus—represents $10 billion or more in capital expenditure, just for computing hardware. Add construction, land, power infrastructure, and contingency, and total investment for a major AI facility reaches $15 billion to $25 billion.

Only a handful of organizations can afford this: the largest technology companies, sovereign wealth funds, and private equity firms with access to cheap capital. This concentration shapes the AI industry. Startups cannot compete on infrastructure; they must rent computing from the giants or accept positions in less compute-intensive niches.

Capital costs are only the beginning. AI hardware is expensive to run.

A B300 GPU draws 1,400 watts under full load. Operating continuously at $0.06 per kilowatt-hour—a favorable rate available only to large industrial customers—each GPU consumes roughly $736 in electricity annually. Add cooling overhead (typically 20 to 30 percent of IT power load) and total power costs reach $1,000 per GPU per year. Maintenance, staffing, and facility overhead add more. For a large deployment, operating costs run into the hundreds of millions annually.

But AI infrastructure economics are dominated by something else entirely: rapid obsolescence. AI hardware loses value with remarkable speed. A GPU purchased in 2023 for $30,000 might be worth $5,000 by 2026, replaced by chips delivering five times the performance at similar prices. Companies depreciate AI hardware over three to five years for accounting purposes, but effective economic life may be only two to three years before performance disadvantages make the equipment uncompetitive.

This obsolescence creates intense pressure to maximize utilization. A GPU that sits idle loses economic value whether it runs or not—the next generation is always coming. Operators aim for 80 to 90 percent utilization, running training jobs continuously and filling gaps with inference workloads. Cloud providers charge premium rates reflecting not just operating costs but the urgency of extracting value before the hardware becomes uncompetitive.

———

Organizations deploying AI at scale face a fundamental choice: build their own infrastructure or rent from cloud providers.

Building offers lower long-term costs for organizations that can maintain high utilization. The break-even point falls somewhere between twelve and twenty-four months of continuous use, depending on costs and utilization rates. Organizations running major training workloads that occupy GPUs for months often find ownership economical.

Renting offers flexibility and speed. Cloud providers maintain inventory and can provision thousands of GPUs within hours. Organizations with variable workloads, uncertain demand, or limited capital prefer renting despite higher hourly costs. Scaling up for major projects and down afterward avoids the risk of stranded assets.

The largest AI companies do both. OpenAI rents substantial capacity from Microsoft Azure while also deploying dedicated hardware. Meta builds its own facilities while maintaining cloud relationships for overflow. xAI built Colossus in Memphis while also using external capacity during the construction period.

This hybrid approach reflects the uncertainty pervading AI infrastructure planning. Nobody knows exactly how much compute future AI systems will require, how quickly chip performance will improve, or how demand for AI services will evolve. Optionality—the ability to shift between owned and rented infrastructure—has significant value even if it results in higher average costs.

The capital intensity of AI infrastructure has attracted sophisticated financial engineering. Companies that cannot fund billion-dollar deployments from operating cash flow have developed creative financing structures.

CoreWeave exemplifies one approach. The company secured billions in debt financing backed by long-term contracts with Microsoft and OpenAI. These contracts provide predictable revenue streams that lenders accept as collateral. The GPU hardware itself provides additional security—though rapidly depreciating assets make challenging collateral. CoreWeave's credit facilities reached over $12 billion by early 2025, terms inconceivable for a startup just two years earlier.

Private equity has entered aggressively. Blackstone, DigitalBridge, and other firms raised dedicated funds for AI infrastructure, viewing data centers as a new asset class comparable to real estate or energy infrastructure. The attractions are clear: long-term contracts with creditworthy tenants, predictable operating costs, demand growth that appears durable. But the risks are substantial: technology obsolescence, tenant concentration, power grid constraints that could limit growth.

Sale-leaseback arrangements allow operators to monetize existing infrastructure while retaining operational control. A company builds a data center, sells the building to an investor, and leases it back under a long-term agreement. This frees capital for expansion while transferring real estate risk to investors with lower cost of capital.

Financing structures continue to evolve. Some companies have explored GPU-backed lending, treating computing hardware as collateral analogous to aircraft or ships. Others have proposed synthetic ownership structures where investors purchase fractional interests in GPU capacity. The creativity reflects desperation: the capital required to compete in AI infrastructure exceeds what traditional financing can easily provide.

Risk profiles vary enormously. A contract with Microsoft backing a CoreWeave deployment carries different risk than a speculative build by a new market entrant. Lenders and investors are still learning to distinguish viable infrastructure plays from overoptimistic projections. The inevitable shakeout will determine which financing approaches survive and which become cautionary tales.

David, a private equity partner at an infrastructure fund, spent the autumn of 2024 trying to understand the chip supply chain. His fund had been investing in data centers since 2009, but the AI transition had changed everything. GPUs were in short supply. Lead times stretched to eighteen months. Secondary market prices doubled. The obvious play was to chase chips.

He built models. He talked to NVIDIA contacts. He tracked allocation tiers and secondary market pricing. The analysis pointed toward chip scarcity as the binding constraint. But something did not fit. Microsoft's Satya Nadella had mentioned chips sitting idle—inventory the company could not "light up." CoreWeave had GPUs but needed power infrastructure. The bottleneck was shifting.

"I spent six weeks chasing the wrong problem," *David* would later say. "Everyone was obsessed with chips. Chips, chips, chips. It took me embarrassingly long to realize: chips are a stock problem. You can accumulate them. Power is a flow problem. You use it or you lose it." The insight would reshape his fund's investment thesis—and send him searching for projects with secured grid access rather than GPU allocations.

2.8 THE SCALE OF DEPLOYMENT

The largest AI deployments now rival power plants in energy consumption. xAI's Colossus cluster in Memphis reached 100,000 NVIDIA Hopper GPUs in 122 days—proving that existing industrial buildings can be converted to AI data centers on startup timelines rather than utility timelines.[52] CoreWeave operates thirty-three facilities with over 250,000 GPUs, each built since 2025 supporting 130-kilowatt rack densities with liquid cooling as standard. Microsoft accounts for 62 percent of CoreWeave's revenue; OpenAI has signed contracts totaling $22.4 billion.[53]

Meta's Prometheus facility in New Albany, Ohio, represents the next frontier: over 500,000 GPUs, power consumption exceeding one gigawatt, investment measured in tens of billions.[54] Unlike competitors deploying almost exclusively NVIDIA hardware, Prometheus will include AMD MI300X accelerators and Meta's custom MTIA chips alongside Blackwell GPUs—hedging against supply chain concentration.[54]

The Stargate initiative announced in January 2025 claimed $500 billion in AI infrastructure investment over four years, with SoftBank, OpenAI, and Oracle as partners.[55] Whether that scale is achievable remains uncertain. Even NVIDIA's expanded production cannot keep pace with projected chip demand. Power grid infrastructure requires years to build. Skilled labor is scarce. The Saline Township project sits within this larger story: one facility among hundreds in the largest private infrastructure buildout in American history.

2.9 THE SILICON ARRIVES IN SALINE

In the spring of 2026, the first trucks carrying computing equipment will arrive at the Saline Township construction site. The data halls will still be under construction—enormous single-story structures rising from what had been soybean fields months earlier. But the infrastructure to receive the chips will be ready: liquid cooling systems, electrical substations, fiber optic connections.

The chips represent the endpoint of everything this chapter has described: Taiwan fabrication, Korean memory packaging, California design, Texas integration—all converging on a former farm in rural Michigan. The initial deployment targets NVIDIA GB200 Grace Blackwell Superchips, with capacity planned for Blackwell Ultra and Rubin-generation systems as they become available.

The silicon in Saline Township embodies all the contradictions of the AI revolution: extraordinary capability enabled by extraordinary consumption, transformative technology dependent on fragile supply chains, private profit built on public infrastructure. Understanding the chips is essential to understanding what is being built. But the chips are only part of the story.

For the residents, the chips will remain abstract—boxes arriving on trucks, disappearing into buildings where visitors are not permitted. What they see is construction: cleared fields, excavation, steel and concrete rising from the earth. They see power lines and substations. They hear about water usage and tax implications. The silicon will do its work invisibly, but the infrastructure required to support it is already transforming land they have known for generations.

Saline Township sits at an intersection: where global technology trends meet local realities, where trillion-dollar supply chains terminate in former agricultural land. The chips inside matter immensely for the future of technology. For the people living nearby, what matters is everything around the chips: power, water, traffic, jobs, taxes, and the transformation of their community into something no one fully anticipated.

The student in Ann Arbor has not thought about silicon. She typed a question, and words appeared. The interface hid everything: *Wei-Lin*'s clean room in Taiwan, the geopolitical tensions that constrain who can buy what chips, the thermal engineering that keeps processors from destroying themselves, the financial structures that make billion-dollar deployments possible.

We have followed her token from mathematics to matter. But the matter itself is precarious. A single policy shift in Washington or Beijing could reshape the entire industry. A single earthquake in Taiwan could halt production of every advanced AI chip in the world. A single breakthrough in efficiency—like DeepSeek's—could strand billions of dollars in infrastructure. The supply chain that delivers her answer is simultaneously the most sophisticated and the most fragile system human beings have ever built.

Silicon alone accomplishes nothing. The chips need buildings. They need cooling systems that can remove heat faster than it accumulates. They need power distribution that delivers electricity without interruption. A chip sitting in a warehouse is idle capital, depreciating by the day. It becomes valuable only when installed in a working data center.

Her token traveled from concept to chip. Now we follow the chips into the buildings that house them.

WHAT COMES NEXT

The chips require massive facilities, cooling systems to prevent them from destroying themselves, physical infrastructure that makes computation possible at scale. Chapter Three examines the data center itself: the architecture of these massive facilities, how they are designed and built, what they cost to construct and operate, how they function as the physical substrate of artificial intelligence. The buildings matter as much as the silicon they contain.

The Data Center

THE CONSTRUCTION SITE STRETCHES across a 575-acre parcel of Washtenaw County farmland, with the skeletal frames of three massive data halls rising against a gray December sky.[56,57] *Steve*, site superintendent for Related Digital, stands beside a muddy pickup truck, gesturing toward the nearest structure. "That's Data Hall One," he says. "When it's finished, those walls will hold more computing power than existed in the entire world twenty years ago."

Steve has spent thirty-two years building things most people never notice. He started framing houses in Phoenix after high school, moved to commercial construction in his twenties, and found his way to data centers in 2008 when a contractor needed someone who could read electrical prints. "Back then, we were building server rooms in office parks," he says. "Twenty kilowatts per rack, maybe thirty. Air conditioning could handle it. A good HVAC crew was all you needed."

He pulls out his phone and scrolls to a photo: the interior of a Facebook data center in Iowa, taken during a 2019 project. Rows of blue-lit servers stretch into the distance like a cathedral. "Beautiful facility. State of the art. Totally obsolete for what we're building now." The Saline Township project requires liquid cooling piped to every rack, power densities that would have seemed impossible five years ago, and construction timelines that keep his crews working twelve-hour shifts six days a week.

The scale is difficult to comprehend. Three data halls, each roughly 550,000 square feet, will house the servers themselves—football-field-sized rooms filled with row upon row of equipment. Separate structures will contain backup generators, electrical substations, and administrative offices. Only about 250 of the 575 acres will be developed; the rest will remain as open space, wetlands, and conservation land. When complete, the campus will consume as much electricity as a city of 800,000 people.

"Most folks drive past data centers every day and never think about what's inside," *Steve* says. "They see a windowless building with some trucks out front. They don't see the engineering that keeps these machines running around the clock, every day of the year."

He points to a trench along the building's perimeter where workers in hard hats are laying thick copper pipes. "Cooling," he explains. "The chips generate so much heat, we can't use air conditioning anymore. We pump liquid directly onto the processors. The whole building is basically one giant radiator."

This is the Related Digital Saline Township campus, one of several mega-facilities being built across Michigan for Oracle and OpenAI as part of the Stargate initiative. When fully operational, it will represent more than seven billion dollars in investment and require roughly 1.4 gigawatts of new power generation from DTE Energy.[18,57,58] It exists because the artificial intelligence systems transforming our economy need physical homes—and those homes have requirements that push the boundaries of modern engineering.

3.1 WHAT IS A DATA CENTER?

At its simplest, a data center is a building designed to house computers. But that definition undersells the complexity. A modern data center is a carefully orchestrated system where power, cooling, networking, and physical security work together to keep servers running continuously. A failure in any of these systems can bring down the entire facility.

The concept is older than most people realize. The first recognizable data center appeared in 1946 at the University of Pennsylvania, where ENIAC—the Electronic Numerical Integrator and Computer—occupied an entire room. That room had to be climate-controlled because ENIAC's eighteen thousand vacuum tubes generated enormous heat. The patterns established then persist today: specialized buildings, climate control, power management. Only the scale has changed.

The internet boom of the 1990s professionalized the industry. Companies like Exodus Communications built facilities designed specifically to house servers, offering space, power, and connectivity to businesses that wanted an internet presence without building their own infrastructure. These colocation facilities—where operators provide space, power, and cooling while customers install their own servers—established practices that persist today: raised floors, hot aisle and cold aisle arrangements, redundant power feeds, standardized rack formats.

The cloud computing era, beginning around 2006 with Amazon Web Services, shifted scale once again. Instead of thousands of companies each running small data centers, a handful of cloud providers built massive facilities serving millions of customers. These hyperscale data centers pioneered innovations in efficiency, automation, and standardization that the rest of the industry eventually adopted.

The core function remains unchanged: computers process information, and processing generates heat. The data center's job is to deliver electricity and remove heat without interruption—simple in principle, extraordinarily difficult at scale. A kilowatt powers a typical hair dryer; a megawatt equals a thousand kilowatts and can power roughly 750 homes; a gigawatt equals a thousand megawatts and requires a large power plant to produce.

The inference operations described in Chapter 1 have a physical cost. Each processor draws electricity and produces heat. AI processors generate heat at rates that air cooling cannot handle. A single AI server rack

produces as much heat as forty household furnaces running simultaneously.

This heat problem drives modern data center design. The buildings
rising in Saline Township look like ordinary warehouses from the outside,
but inside they contain some of the most sophisticated thermal management systems ever built. Understanding those systems starts with the
basic unit of data center organization: the rack.

3.2 THE RACK AS BUILDING BLOCK

Walk into any data center and you encounter rows of identical black cabinets, each about seven feet tall, two feet wide, and three feet deep. Of
course, you cannot actually walk into any data center—attempting to do
so would most likely end with armed security escorting you off the property, or with local police responding to a trespassing report. But suspend
disbelief for a moment and imagine you found your way past the fencing, the biometric locks, and the guards. These server racks provide the
organizing framework for everything that happens inside.

A standard rack contains 42 "units" of vertical space, each 1.75 inches
tall. Server manufacturers design their equipment to fit these units. A
small server might occupy one unit; a larger system might require two
or four. This standardized form factor allows equipment from different
manufacturers to work together in the same rack.

The rack provides more than physical structure. It delivers power
through vertical strips of electrical outlets. It routes network cables to
switches connecting each server to the rest of the facility. It manages
airflow, directing cool air to the front of the servers and exhausting hot
air from the back. In modern AI data centers, it also distributes liquid
cooling through pipes running the height of the cabinet.

What has changed dramatically is power consumption. In 2010, a typical rack drew five to fifteen kilowatts. The servers inside were general-
purpose machines running websites, email systems, and business appli-

cations. They generated heat, certainly, but nothing a well-designed air cooling system could not handle.

By 2020, cloud computing had pushed rack densities higher. Virtualization allowed multiple applications to share the same physical hardware, making servers work harder and draw more power. Typical racks in cloud data centers consumed fifteen to thirty kilowatts. Air cooling still worked, but operators had to manage airflow more carefully, often installing barriers to prevent hot exhaust from mixing with cool intake air.

Then AI arrived. NVIDIA released its A100 processor in 2020, designed for training large AI models. A server containing eight A100 chips drew fifteen to twenty kilowatts by itself. Fill a rack with several such servers, and consumption approached sixty to eighty kilowatts. By 2024, the H100 pushed power draw higher still, with racks consuming a hundred kilowatts or more.

The newest hardware takes another leap. The GB300 NVL72 systems described in Chapter 2—rack-scale configurations drawing 140 kilowatts—produce enough heat to warm several houses through a Michigan winter.

This sevenfold increase in power density over fifteen years has transformed data center engineering. Facilities designed for traditional workloads cannot accommodate AI equipment without major retrofits. The buildings rising in Saline Township are purpose-built for this reality.

3.3 INSIDE THE SERVER

The rack is just a container. The real work happens inside the servers—the individual computers filling each rack. Understanding what makes AI servers different from their predecessors helps explain why the entire data center has had to change around them.

A traditional server resembles a pizza box, one to two units tall, containing a motherboard with one or two general-purpose CPUs, memory slots, storage drives, and networking ports. These machines excel at diverse workloads: running web applications, processing database queries, handling email. The CPUs inside are optimized for flexibility, capable of executing almost any software efficiently.

AI servers look different. They are larger—four to eight units tall—and contain specialized processors alongside traditional CPUs. The most common AI accelerators are NVIDIA GPUs, graphics processing units originally designed for rendering video games but now repurposed for the parallel mathematical operations that AI requires. Alternatives from AMD, Intel, and custom designs from Google and Amazon also see deployment. These accelerators sacrifice the flexibility of general-purpose CPUs for raw mathematical throughput. A single H100 GPU can perform orders of magnitude more floating-point operations per second than a high-end CPU, but only for the specific calculations that machine learning requires.

What GPUs need, above all, is memory bandwidth. Training an AI model involves moving vast quantities of data through the processors—terabytes per second in a large system. Traditional computer memory cannot feed data to GPUs quickly enough. High Bandwidth Memory, or HBM, addresses this bottleneck by stacking memory chips vertically and connecting them to processors through thousands of parallel pathways. An H100 GPU includes 80 gigabytes of HBM with over three terabytes per second of bandwidth, roughly ten times faster than conventional memory.

The servers also contain high-speed interconnects linking GPUs together. Within a single server, NVIDIA's NVLink allows eight GPUs to share data at 1.8 terabytes per second. This internal bandwidth lets the GPUs operate cooperatively, splitting large calculations across all available processors. Without these interconnects, each GPU would work in isolation, unable to collaborate on problems too large for any single chip to solve.

Networking differs too. Standard gigabit Ethernet, common in older data centers, cannot handle the traffic that AI training generates. AI servers typically include multiple ports running at one hundred or two hundred gigabits per second, using either InfiniBand or high-speed Ethernet with RDMA support. These ports connect to the facility's network fabric, allowing processors in different servers and racks to communicate during training runs.

Storage is almost an afterthought. AI training reads from massive datasets that typically live on separate storage systems, streaming training examples to the GPUs as needed. The servers contain only enough local storage for the operating system and temporary files. The real storage infrastructure sits elsewhere in the facility, optimized for the sequential reads that training demands.

Assembling these components into working servers requires extraordinary precision. Each GPU must make contact with its memory through hundreds of solder balls smaller than a grain of sand. Cooling systems must deliver liquid to cold plates machined to tolerances of hundredths of a millimeter. Power delivery must handle rapid fluctuations as workloads vary, without voltage droops that could cause errors. Manufacturing these systems is as much craft as industrial process.

When a server arrives at Saline Township, it undergoes extensive testing before deployment. Technicians verify that all GPUs function, that cooling connections are leak-free, that networking reaches full speed. The machines are expensive—a single server with eight H100 GPUs costs roughly $300,000—and finding problems early avoids costly downtime later. Once deployed, monitoring systems track hundreds of parameters per server, alerting operators to developing problems before they cause failures.

3.4 POWER: THE LIFEBLOOD

Before a data center can do anything else, it must have electricity. Power infrastructure accounts for much of a facility's cost and complexity, and for good reason: even a brief interruption can corrupt an AI model that has been training for weeks.

Electricity arrives through utility connections, typically at high voltage. The Saline Township campus will connect to DTE Energy's transmission network, drawing power from a mix of natural gas plants, nuclear reactors, and renewable sources. That incoming power must be converted to lower voltages that servers can use, and protected against any interruption in utility supply.

Conversion happens through a series of transformers and power distribution units. High-voltage transmission power—often 138,000 volts or more—steps down to medium voltage, then to the 415-volt three-phase power—industrial-grade delivery using three synchronized alternating currents—that modern server racks require. Each step involves equipment that generates heat and introduces potential points of failure.

Protection against power interruption involves two complementary systems. Uninterruptible power supplies, or UPS systems, contain batteries that can take over instantly if utility power fails. These batteries cannot run the facility for long—typically fifteen to thirty minutes—but they bridge the gap while backup generators start.

Generators provide the second layer of protection. Data centers maintain fleets of diesel or natural gas generators capable of powering the entire facility indefinitely. The generator building at Saline Township will house dozens of these machines, each the size of a shipping container. When utility power fails, UPS batteries keep servers running while generators spin up. Within a minute or two, generator power takes over, and the facility operates independently of the grid as long as fuel lasts.

The industry describes this redundancy using "N" notation. A facility with N+1 power redundancy has one more power path than it needs to operate at full capacity. If any single component fails, the remaining equipment handles the load. More critical facilities implement 2N redundancy, meaning the entire power infrastructure is duplicated. Either half can run the facility alone, allowing maintenance on one set of equipment without risking operations.

Power distribution follows a hierarchical structure. Main switchgear receives power from both the utility and generators. From there, power flows to electrical rooms distributed throughout the building, then to power distribution units mounted in or near the server racks. At each level, circuit breakers and monitoring equipment protect against faults and provide visibility into consumption.

The cost is substantial. A megawatt of power capacity—enough to run about eight hundred homes at any given moment—might cost one to two million dollars to install, accounting for transformers, switchgear, UPS systems, generators, and distribution equipment. The Saline Township campus, with its 1.4-gigawatt eventual capacity, represents billions of dollars in power infrastructure alone.

3.5 COOLING: THE THERMAL CHALLENGE

"You want to know the secret of this business?" *Steve* says, watching his crew lower a coolant distribution unit into place. "It's not computers. It's plumbing." He laughs, but he is not joking. In thirty-two years of construction, he has watched data centers transform from buildings that happened to need air conditioning into buildings that are fundamentally heat-removal systems with some computing attached.

Why does computing generate heat? The answer lies in basic physics. When electrons flow through circuits, they encounter resistance—atoms that slow their passage through copper and silicon. Each collision converts a tiny bit of electrical energy into thermal energy. This is the same principle that makes incandescent light bulbs glow and toaster coils turn red. Energy is never destroyed, only transformed. In a data center, all the electricity that enters eventually becomes heat. There is no exception.

Steve learned this the hard way on a job in 2011, when a cooling system failed and a server room hit 140 degrees in twenty minutes. "Smelled like burning plastic and money," he says. "Three million dollars of equipment, cooked. After that, I started paying attention in the mechanical briefings."

Every watt of electricity entering a data center leaves as heat. A facility consuming a gigawatt of power produces a gigawatt of heat that must be removed.

Traditional data centers addressed this with air cooling—cold air flowing through racks, absorbing heat, exhausting outside. The approach worked for decades at densities below twenty-five or thirty kilowatts per rack. But physics sets hard limits. Above thirty kilowatts, the volume of air required becomes impractical. The fans consume significant power themselves and generate noise approaching dangerous levels. When CoreWeave, xAI, and Meta started deploying AI clusters at a hundred kilowatts per rack, they had no choice but to adopt liquid cooling.

Liquid cooling comes in several forms, but the most common for AI data centers is direct-to-chip cooling. Cold plates—precisely machined metal blocks with internal channels—mount directly on processors and other high-power components. Coolant flows through these plates, absorbing heat at the source before it can spread to the surrounding air. The warmed coolant travels to a heat exchanger, transferring its heat to the facility's chilled water system or directly to outdoor cooling towers.

The efficiency gains are dramatic. CoreWeave reports that direct-to-chip cooling removes eighty percent or more of the heat from AI racks at the component level.[59,60] The remaining heat from lower-power components like network cards and storage drives can be handled by supplementary air cooling. Supermicro's DLC-2 system claims ninety-eight percent heat capture at the chip level, leaving almost nothing for air systems.

This efficiency translates to operational savings. Liquid-cooled facilities can operate at power usage effectiveness ratios—PUE, the metric comparing total facility power to IT equipment power—of 1.2 to 1.3. For every megawatt of computing power, the facility needs only 0.2 to 0.3

megawatts for cooling and other overhead. Traditional air-cooled facilities often run at PUE values of 1.6 to 1.8, with nearly half their power going to cooling.

Liquid cooling adds complexity and cost. Coolant Distribution Units, or CDUs, serve as the interface between the facility's chilled water and the server-level cooling loops. A single CDU might support one to four racks, depending on power density. The Saline Township facility will contain hundreds of these units, each requiring connections to both the facility chilled water loop and the server coolant manifolds.

Server-level coolant loops use quick-disconnect couplings that allow hot-swapping equipment without draining the entire system. When a server fails, technicians can disconnect the coolant lines, remove the server, install a replacement, and reconnect—all without interrupting the rest of the rack. The precision engineering required for these connections, which must handle pressures of twenty to forty PSI while preventing leaks, adds significantly to equipment costs.

Beyond direct-to-chip cooling lies immersion cooling, where servers are submerged entirely in tanks of dielectric fluid. This approach can handle densities of two hundred to three hundred kilowatts per rack or higher, absorbing heat across every component surface rather than just the main processors. Companies like GRC and LiquidStack have commercialized immersion systems, and some hyperscalers are experimenting with them for their highest-density deployments.

Immersion cooling offers remarkable efficiency—PUE values as low as 1.1 to 1.2—but introduces operational challenges. Servers must be designed or modified for immersion compatibility. Maintenance requires lifting equipment from fluid-filled tanks, a messier process than working with air-cooled or direct-to-chip systems. The dielectric fluids cost thirty to sixty dollars per gallon and require periodic replacement. For most current AI deployments, direct-to-chip cooling strikes the right balance of performance and practicality.

Cooling generates its own resource demands. Traditional data centers using evaporative cooling towers consume staggering amounts of water—typically one to two million gallons per megawatt per year.[61,62] A large facility might draw as much water as a small town. This invisible demand has begun attracting regulatory scrutiny and sparking community opposition.

The physics are straightforward. Evaporative cooling works by allowing water to evaporate, carrying heat away as it changes from liquid to vapor. The process is thermodynamically efficient, achieving cooling capacities that dry systems struggle to match. But the water is consumed rather than recycled, lost to the atmosphere as vapor.

In water-stressed regions, this consumption has become politically charged. Tucson, Arizona unanimously rejected a proposed Amazon data center in August 2025, citing water concerns.[63] Communities in Phoenix and Las Vegas have questioned whether data centers should receive water allocations that might otherwise serve residents. The conflict between data center growth and water scarcity is sharpening across the American Southwest.

Liquid cooling offers partial relief. Because direct-to-chip systems remove heat at the source with high efficiency, they reduce the total cooling load that must be rejected to the environment. A liquid-cooled facility might need only half the evaporative capacity of a comparable air-cooled facility, cutting water consumption proportionally.

Some operators have gone further, eliminating evaporative cooling entirely. Applied Digital's facility in Ellendale, North Dakota uses closed-loop, waterless, direct-to-chip cooling.[64] The system rejects heat through dry coolers—essentially large radiators that transfer heat to the air without evaporating water. This works well in cold climates, where ambient temperatures allow dry cooling to be effective. In hot climates, dry cooling requires larger equipment and consumes more electricity, making the economics less favorable.

The Saline Township facility will likely use a hybrid approach. Michigan's climate allows significant dry cooling during cooler months, reducing water consumption compared to warmer regions. But summer peaks will still require some evaporative cooling to maintain safe operating temperatures. The consent agreement governing the project includes provisions for monitoring and mitigating impacts on neighboring wells and groundwater—an acknowledgment that water consumption affects the surrounding community.[5,65]

Heat reuse represents another frontier. Liquid cooling captures heat at temperatures high enough to be useful—forty to sixty degrees Celsius in some systems. That warm water could heat nearby buildings, displacing fossil fuel consumption. Several European data centers already sell waste heat to district heating networks.[66]

Sixty-five miles northwest of Saline Township, a different kind of data center project is taking shape. Deep Green, a UK-based company, proposed a 24-megawatt facility on a downtown Lansing parking lot. The site sits in the city's stadium district, zoned for downtown core development, not industry. The company needs City Council to approve a conditional rezoning before construction can begin. The contrast with Saline could not be sharper: instead of lawsuit and settlement, an open house and a public vote scheduled for February 2026. Instead of 1.4 gigawatts serving a single hyperscaler, 24 megawatts serving universities running physics simulations, studios rendering CGI effects, and pharmaceutical companies doing drug discovery. Instead of wasted heat, a donation to the Board of Water and Light, Lansing's municipally-owned utility, which plans to use it to replace the city's century-old steam heating system with a modern hot water network.

The Deep Green model works because the scale is manageable. Twenty-four megawatts fits on a parking lot. The heat output matches a city's district heating needs. The power demand does not overwhelm the local grid. At hyperscale, these arrangements become impractical. A facility consuming 1.4 gigawatts produces enough waste heat to warm a city of hundreds of thousands, but no American city has district heating infrastructure at that scale. The heat dissipates into the atmosphere, a thermodynamic necessity that smaller European-style facilities can avoid.

3.6 Network: The Nervous System

Computing power without connectivity is useless. Data centers must connect to the broader internet to serve users, and they must link their internal systems so processors can collaborate on large tasks. Networking in an AI facility differs substantially from traditional data center networking, optimized for the communication patterns unique to machine learning.

External connectivity comes through fiber optic cables linking the data center to internet exchange points and content delivery networks. A facility like Saline Township might connect to multiple regional exchanges and peer directly with internet service providers. These connections must be redundant—if one fiber route fails, traffic reroutes automatically through alternatives.

The more interesting challenge is internal. Training a large AI model requires thousands of processors working in concert, constantly exchanging data as they refine the model's parameters. The speed of these interconnections often determines how quickly training completes. If processors spend too much time waiting for data from their neighbors, expensive hardware sits idle.

Traditional Ethernet networking, designed for general-purpose traffic, struggles with AI workloads. The protocol introduces latency as it processes packets, and its congestion management algorithms were designed for diverse traffic patterns rather than the all-to-all communication that AI training requires.

Two alternatives have emerged. NVIDIA's InfiniBand technology, originally developed for supercomputing, delivers higher bandwidth and lower latency than Ethernet. The company's Quantum-2 switches provide 400 gigabits per second per port, with latency measured in microseconds rather than the hundreds typical of Ethernet switches.

The alternative is RDMA over Converged Ethernet, or RoCE, which adapts Remote Direct Memory Access—the same underlying technology as InfiniBand—to run over standard Ethernet switches. RoCE offers lower hardware costs and familiarity for operators accustomed to Ethernet. xAI chose this approach for its Colossus supercomputer, connecting a hundred thousand H100 processors over a single RoCE fabric.[52]

Within a rack, NVIDIA's NVLink provides even faster connections between processors. The high-speed interconnects described in Chapter 2 allow processors to operate almost as a single massive chip, sharing memory and coordinating computations at extraordinary speeds.

Network topology—how switches and servers connect—affects both performance and reliability. Most AI data centers use a "fat tree" or "spine-leaf" architecture: top-of-rack switches connect servers within each rack, leaf switches aggregate traffic from multiple racks, and spine switches interconnect the leaf layer. This design provides multiple paths between any two points, allowing traffic to route around failures and preventing any single switch from becoming a bottleneck.

3.7 SECURITY: PHYSICAL AND DIGITAL

Data centers house valuable assets: expensive equipment and, more importantly, the data and services running on it. Security involves both physical protection—preventing unauthorized entry—and operational security—controlling what authorized personnel can access and do.

Physical security begins at the property boundary. The Saline Township site will feature fencing, surveillance cameras, and controlled vehicle access. Visitors pass through a gatehouse where guards verify credentials against an approved list. Once past the perimeter, additional barriers separate the facility into security zones, with increasingly restrictive access as you approach the server halls.

Entry to the building requires authentication, typically badge access combined with biometric verification. Modern facilities use fingerprint or iris scanners to ensure only the badge holder can use their credentials. Once inside, staff can access only the areas their role requires. A facilities technician might enter mechanical rooms but not server halls. An IT technician might access certain rows of racks but not others.

Within server halls, surveillance cameras record all activity. Most facilities implement a "man trap" at entry points—an airlock-like vestibule where one door must close before the other opens, preventing tailgating. Some require two-person integrity for sensitive areas, meaning no one can access those spaces alone.

These precautions might seem excessive for buildings that house computers. But the services running on those machines can be extraordinarily valuable. A disruption to AI training that has run for weeks costs not just the electricity consumed but the time lost. If the weights of a leading AI model were stolen, competitors could replicate months or years of research investment. National security concerns add another dimension: AI systems developed at facilities like Saline Township have implications for military and intelligence applications.

Operational security governs who can access what within the digital systems. Not everyone with physical access to a rack should be able to access the data it processes. OpenAI, as a tenant at Saline, will control what runs on its leased equipment, with Related Digital providing the physical infrastructure but not accessing the computing environment. This separation—facilities management distinct from computing operations— is common in colocation arrangements.

3.8 WHEN THINGS GO WRONG

For all the redundancy built into modern data centers, failures still occur. Understanding what can go wrong—and how facilities survive failures—reveals the engineering philosophy behind these buildings.

Power failures represent the most common catastrophic risk. Utility outages happen: storms damage transmission lines, equipment fails at substations, grid operators shed load during emergencies. UPS and generator systems exist for precisely these moments. When utility power disappears, the UPS takes over within milliseconds. Generators start automatically, reaching operating speed within thirty seconds to two minutes. Once generator power stabilizes, the system transfers load from batteries to generators, which can run indefinitely with fuel.

The transfer between power sources is the moment of greatest vulnerability. If the UPS fails to engage, servers lose power instantly. If generators fail to start, batteries eventually drain. If the transfer introduces voltage anomalies, sensitive electronics can malfunction. Engineers design and test these systems exhaustively, but failures still occur. In 2017, an outage at British Airways' data center caused by power system failure cost the airline an estimated eighty million pounds in cancelled flights and compensation.[67]

Cooling failures can be equally devastating. Servers generate heat continuously, and without cooling, temperatures rise dangerously within minutes. A complete cooling failure gives operators perhaps fifteen to thirty minutes before equipment begins shutting down due to thermal protection. In liquid-cooled facilities, a major leak could damage multiple servers before containment systems respond. Data centers implement multiple layers of cooling redundancy, but cascade failures—where one problem triggers others—remain possible.

Human error causes a surprising fraction of incidents. A technician disconnecting the wrong cable can take down an entire network segment. A maintenance procedure that accidentally trips a circuit breaker can cut power to critical systems. Even mundane actions—spilling a drink, dropping a tool—have caused outages. Procedures, training, and careful design reduce these risks but never eliminate them.

Natural disasters pose site-specific risks. Facilities in tornado-prone regions must withstand high winds. Those near coastlines must consider flooding and storm surge. Earthquake zones require seismic design. The Saline Township site, in southern Michigan, faces relatively modest natural hazard risks—occasional severe thunderstorms and winter weather, but no significant earthquake or hurricane exposure.

Achieving five nines reliability requires not just redundant systems but rigorous maintenance, thorough procedures, and a culture that treats even minor incidents as opportunities for improvement. The most reliable facilities document every failure, investigate root causes, and implement changes to prevent recurrence.

3.9 THE PEOPLE INSIDE

Behind all this infrastructure are the people who keep it running. A large data center employs hundreds across multiple disciplines: facilities engineers maintaining power and cooling, security guards controlling access, network technicians troubleshooting connectivity.

Operations runs around the clock. The systems that keep servers running cannot take nights or weekends off, and problems can develop at any hour. A typical shift might include a chief engineer, several facilities technicians, a network operations specialist, and security personnel. They monitor thousands of sensors tracking power consumption, temperatures, humidity, and equipment status, responding to alerts that could indicate developing problems.

The skills required have evolved with the technology. Traditional data centers needed expertise in electrical systems, air conditioning, and standard IT networking. AI facilities add liquid cooling to the mix. Technicians must understand coolant chemistry, manage pressure and flow rates in complex piping systems, and handle the quick-disconnect couplings that allow hot-swapping liquid-cooled servers.

Finding people with these skills has become difficult. The rapid build-out of AI infrastructure has created demand that the labor market struggles to meet. Companies compete to hire experienced facilities engineers, sometimes offering substantial premiums over traditional data center roles. Training programs are emerging to develop this workforce, but the pipeline cannot keep pace with the hundreds of facilities under construction or planned.

The work is physically demanding. Server equipment is heavy, and repairs may require lifting components weighing fifty pounds or more. Server halls are loud—constant fan noise layered over mechanical systems operating throughout the facility. Temperatures vary dramatically between cold aisles and hot aisles, sometimes differing by thirty or forty degrees Fahrenheit within a few feet.

Safety training is extensive. Electrical systems in a data center can deliver lethal currents. Arc flash incidents—explosive releases of energy from electrical faults—represent a constant hazard that requires careful procedures when working on live equipment. Liquid cooling introduces leak risks that could damage expensive hardware or create slip hazards.

Despite these challenges, many facilities professionals find the work rewarding. They build and maintain infrastructure that powers technologies transforming society. The AI systems trained in these buildings will influence medicine, science, transportation, and countless other fields. That sense of purpose helps attract and retain talent in a competitive market.

This brief survey understates what the labor story deserves. The workforce building America's AI infrastructure—the electricians who wire these facilities, the pipefitters who install cooling systems, the construction crews working twelve-hour shifts—merit sustained attention that this book does not provide. Their experiences, their negotiations with developers and unions, their career disruptions and opportunities: these are consequential stories unfolding alongside the political economy we examine here.

Beyond the people inside these facilities, millions of workers now depend on what runs here. Software developers use AI coding assistants. Researchers run experiments on rented GPU clusters. Customer service teams rely on AI chatbots. The facilities employ hundreds. The systems they run serve hundreds of millions.

3.10 THE SUPPLY CHAIN

Building a data center requires equipment from dozens of manufacturers across multiple continents. Transformers come from one supplier, switchgear from another, generators from a third. Servers might be assembled in Texas from components manufactured in Taiwan, South Korea, and China. Cooling systems incorporate compressors, pumps, heat exchangers, and piping from yet more sources. Coordinating this global supply chain is as challenging as any other aspect of development.

The COVID-19 pandemic exposed vulnerabilities in these supply chains. Lead times for electrical equipment extended from months to years. Transformer deliveries that once took twelve months stretched to thirty-six or more. Server components faced shortages as pandemic-induced demand for computing coincided with manufacturing disruptions. Delays cascaded through project schedules, pushing facility openings later and increasing costs.

The AI boom has created new bottlenecks. The chip supply constraints described in Chapter 2—Taiwan fabrication, allocation hierarchies, geopolitical tensions—apply equally to the cooling components these facilities require. The most sought-after equipment—particularly the specialized quick-disconnect couplings for liquid cooling—faces similar constraints. Suppliers have expanded capacity, but catching up to exponentially growing demand takes time.

Developers have adapted by ordering equipment earlier, maintaining larger inventories, and qualifying multiple suppliers for critical components. The largest hyperscalers command enough purchasing power to secure priority allocation from manufacturers. Smaller operators must be more creative, sometimes redesigning systems to use available components rather than waiting for preferred alternatives.

For Saline Township, supply chain management began years before construction started. Major electrical equipment was ordered in 2023 for installation in 2025 and 2026. Server orders were placed before the facility design was complete, locking in allocation from NVIDIA and server manufacturers. The project schedule incorporates buffers for potential delays, but supply chain disruptions could still affect timelines.

Moving equipment to site presents its own challenges. Transformers large enough to serve a gigawatt facility weigh hundreds of tons and require specialized transport. Roads must be assessed for weight limits and clearance heights. Permits must be obtained for oversize loads. Scheduling coordinates with utility companies, trucking firms, and local authorities. A single transformer delivery might involve months of planning.

Saline Township illustrates how modern AI facilities are structured. Related Digital, a subsidiary of real estate developer Related Companies, is building the physical infrastructure. Oracle will operate the computing environment, providing AI infrastructure as a service. OpenAI is a major tenant, using that infrastructure to train its AI systems. This layered structure allows each party to focus on its core competency: Related on real estate and facilities, Oracle on cloud infrastructure, OpenAI on AI research.

3.11 THE ECONOMICS OF SCALE

Building a data center is extraordinarily expensive. Saline Township will cost approximately seven billion dollars before a single server runs its first calculation. That covers land, buildings, power infrastructure, cooling systems, networking, and site preparation. Servers themselves? Additional billions from tenants.

Why do data centers keep getting larger? Fixed costs. A facility needs switchgear whether it serves a hundred megawatts or a thousand. Security perimeters cost nearly the same for small sites as large ones. Engineering teams must be hired regardless of scale. Spreading these fixed costs across more computing capacity reduces cost per megawatt.

Power infrastructure typically accounts for thirty to forty percent of construction costs: utility interconnection, transformers, switchgear, UPS systems, generators, distribution equipment. Scale brings bargaining power. A one-gigawatt campus commands attention from vendors that a ten-megawatt facility never will.

Cooling adds another twenty to thirty percent. Liquid cooling costs more than traditional air systems upfront. But it enables the high densities that make AI workloads viable, and lower operating costs offset the premium over time.

The building shell and site work account for much of the remainder. Data centers require specialized construction: raised floors for cable management, reinforced structures to support heavy equipment, electromagnetic shielding in some cases. Site preparation can be substantial, particularly for large campuses on undeveloped land. Saline Township required grading hundreds of acres, installing drainage systems, and running utilities to a site that was previously farmland.

Operating costs accumulate over time. Electricity is the largest expense, often exceeding original construction cost over a facility's twenty-year lifespan. A gigawatt data center running at typical utilization consumes roughly eight billion kilowatt-hours per year.[61] At industrial electricity rates, that translates to hundreds of millions of dollars annually. Staffing, maintenance, and property taxes add further ongoing expenses.

The economics work only if the facility stays utilized. An empty data center still draws power for cooling and basic systems, and still requires staff for security and maintenance. Fixed costs continue even when no revenue-generating workloads run on the equipment. This creates pressure to secure long-term tenant commitments before construction begins—precisely the model Related Digital used with Oracle and OpenAI at Saline Township.

For AI workloads, the economics have unique characteristics. Training runs consume enormous resources for weeks or months, then complete. Inference—running trained models to serve users—requires capacity that grows with demand. The mix between training and inference affects facility design and utilization patterns. A facility optimized for training might look different from one designed for inference at scale.

The capital intensity has attracted substantial financial investment. Private equity firms like Blackstone and DigitalBridge have deployed tens of billions into data center assets. These financial sponsors bring capital and development expertise, partnering with operators who bring technical knowledge. The result is a rapid buildout that neither technology companies nor real estate developers could achieve alone.

———

On a cold January morning, construction proceeds on multiple fronts. The first data hall has reached interior work; the second remains a steel frame exposed to the elements. Workers pour concrete for the generator building as excavators prepare the third data hall site.

Inside Data Hall One, the scale becomes real. The interior stretches longer than a football field, ceilings forty feet high. Blue-painted coolant pipes run along the ceiling, branching down where CDUs will mount. Workers install quick-connect fittings that will let individual racks plug into the cooling system.

"Traditional data centers, you run some ductwork, maybe put in some in-row cooling units," says *Maria*, the mechanical engineer overseeing cooling installation. "This is more like building a chemical plant. Thousands of connection points, all handling pressurized coolant, all needing to deliver precise flow rates. The tolerance for error is basically zero."

David visited a similar construction site in Virginia the previous summer. He had toured dozens of data centers over his fifteen-year career—had walked past racks, peered at cooling systems, nodded along as operators explained their redundancy schemes. But walking a construction site was different. The scale became real in a way the finished facilities obscured.

He watched *Steve*—or someone like *Steve*—manage a crew running liquid cooling pipes through the skeleton of a data hall. The superintendent was explaining something about flow rates and pressure tolerances. *David* took notes, though he knew the technical details would blur together later. What stayed with him was the precision. Every joint, every coupling, every weld had to be perfect. A single leak could destroy millions of dollars in equipment.

"I thought I understood data centers," *David* told his partners afterward. "I understood spreadsheets. I did not understand what it takes to actually build one of these things." The gap between financial models and physical reality had never seemed so wide. He started pricing execution risk differently after that tour.

The electrical infrastructure is designed for 2N redundancy—everything duplicated. Two utility feeds from different substations. Two sets of switchgear. Two independent power paths to every rack. Outside, the generator building will house more than forty units capable of powering a small city. "Forty-two generators," *Steve* says, nodding toward the building. "Natural gas from a dedicated pipeline. This place can run off-grid indefinitely."

"We expect about two hundred people working on site when the first phase is operational," says *Steve*. "Facilities engineers, network technicians, security, logistics people handling the constant flow of equipment and supplies."

Construction schedules call for the first data hall to be operational in late 2026, with additional capacity coming online through 2028 and beyond. The total buildout will take years, representing one of the largest construction projects in Michigan history. When complete, the campus will transform this corner of Washtenaw County, drawing enough power to affect the regional grid and generating enough economic activity to reshape local tax revenues.

Data center construction presents a paradox: facilities built over eighteen to thirty-six months will house equipment that becomes obsolete within five to seven years. The building operates for decades; the computing equipment inside cycles through multiple generations during that time, with each refresh requiring new cooling and power configurations. Some operators compress timelines dramatically—xAI built its Memphis facility in roughly four months—but most balance urgency with prudence, moving as quickly as supply chains and labor allow while building infrastructure designed to outlast multiple generations of computing equipment.

3.12 WHAT MAKES AI DIFFERENT

The preceding sections have described power, cooling, networking, and security as separate systems. AI infrastructure integrates them in ways traditional data centers never required.

The concentration of computing power—the density that drove the shift from air to liquid cooling, the rack evolution from fifteen to 140 kilowatts—creates interdependencies that compound failure risks. When a cooling system fails in an air-cooled facility, operators have thirty to sixty minutes before equipment overheats. In a liquid-cooled AI facility, thermal runaway can occur in minutes. The margin for error shrinks as density increases.

Reliability requirements differ correspondingly. A web server that crashes simply restarts and begins serving requests again. AI training runs for days or weeks, and an interruption can corrupt the entire job. The model must restart from the last saved checkpoint, losing hours or days of progress. This drives the extreme redundancy of AI facility power systems and the emphasis on cooling reliability that prevents thermal shutdowns.

The systems described in this chapter—power, cooling, networking, security—exist in service of the silicon described in Chapter 2. Neither makes sense without the other. A chip without a building is an expensive paperweight. A building without chips is an expensive warehouse. The infrastructure buildout this book examines requires both, at scales and speeds that strain every link in the chain.

This integration challenge explains why hyperscalers—Microsoft, Amazon, Google—hold such asymmetric power in the current moment. They spent the past fifteen years assembling the pieces: land banks near transmission capacity, long-term utility relationships, supplier contracts with favorable allocation, workforces trained in operating complex facilities. A startup with unlimited capital cannot replicate these advantages quickly. Transformers take three years to deliver. Utility interconnection studies take eighteen months. Permits require navigating regulatory processes that reward incumbents with established relationships. The hyperscalers are not merely rich; they are early. In a market defined by lead times measured in years, being early confers structural advantages that money alone cannot buy. Newcomers like CoreWeave and xAI have moved aggressively, but they remain dependent on hyperscaler infrastructure, hyperscaler chip allocations, and sites the hyperscalers passed over.

3.13 EVOLUTION AND FUTURE

The data centers being built today will likely operate for twenty years or more. Over that span, the technology they house will change dramatically. Facilities must accommodate this evolution while continuing to deliver reliable service.

The trend toward higher rack densities shows no sign of slowing. NVIDIA's roadmap suggests processor power consumption will continue increasing, potentially reaching two kilowatts per chip or higher in coming generations. Racks consuming two hundred or three hundred kilowatts may become common within five years. Facilities designed for 140-kilowatt racks will need upgrades to handle these higher densities.

This is why operators like Related Digital build infrastructure with growth in mind. Cooling capacity at Saline Township exceeds what current equipment requires, providing headroom for denser future deployments. Power distribution can accommodate higher loads with relatively modest upgrades. The fiber infrastructure supports bandwidth requirements well beyond what current equipment uses.

Some observers predict architectural shifts beyond the current rack-based design. As densities climb, the rack itself may become limiting. Pod-based designs—clusters of five to ten racks treated as single units—could simplify infrastructure by consolidating power and cooling distribution. Rack-scale systems like NVIDIA's NVL72 already treat an entire rack as a single computer, blurring the traditional distinction between server and rack.

Cooling technology continues advancing. Immersion cooling, currently used in specialized applications, may become mainstream as densities push beyond what direct-to-chip systems can handle. Some researchers explore integrating microfluidic cooling channels directly into chip packaging, removing heat even closer to its source. These technologies remain immature, but the relentless increase in processor power consumption will drive their development.

The facilities being constructed today are bets on the future of artificial intelligence. If AI continues advancing rapidly, demand for infrastructure will only grow. Saline Township and projects like it will provide the physical foundation. If AI development plateaus, these massive investments may prove difficult to recoup.

———

Most people will never see inside a facility like the one rising in Saline Township. The buildings are intentionally anonymous, designed to blend into industrial parks. But step through the airlock of an operational AI data center—the first door closing before the second opens—and the transition is immediate.

The first impression is sound: the constant drone of thousands of fans, not deafening but impossible to ignore. Workers wear hearing protection during extended tasks. The second impression is order. Racks stand in precise rows, their alignment measured in millimeters. Overhead, cable trays carry neatly bundled fiber optic lines, color-coded by function. Nothing here is accidental.

The racks contain servers unlike anything in a traditional data center. Each bristles with cooling connections—the quick-disconnect fittings linking to the building's liquid cooling system. LEDs blink in patterns meaningful to those who know how to read them. Between rows, CDUs hum as pumps circulate coolant. At the end of each row, electrical distribution equipment feeds power through cables thicker than a person's arm.

None of this is visible to users. A researcher training a model sees only a software interface. The physical reality—the careful engineering, the constant maintenance, the orchestration of power and cooling and networking—exists only to make that interface work reliably.

This infrastructure shapes daily life in ways easy to overlook. The AI systems trained in these buildings power recommendation algorithms, voice assistants, search results. They accelerate drug discovery and weather prediction. The chips generating heat in those racks are learning patterns from data that will inform decisions affecting billions of lives.

The construction crew at Saline Township will work through winter, racing to meet schedules set by customers eager to deploy their hardware. By this time next year, the first data hall will be operational, its racks filled with processors, its cooling systems maintaining the precise temperatures that keep those processors running. The farmland that once grew corn and soybeans will host computing power unimaginable a generation ago.

Steve, standing beside the rising structure, looks out over the site. "People ask me what we're building," he says. "I tell them it's a building full of computers. That's the simple answer. The real answer is that we're building a piece of the future."

———

The student in Ann Arbor is still waiting. Not long—milliseconds, though the full response will take a few seconds to stream. The cursor blinks on a laptop screen in a coffee shop on Liberty Street while, fifteen miles southwest, electricity flows through copper bus bars thicker than a human arm. Coolant circulates through pipes machined to tolerances measured in hundredths of a millimeter. Fans spin. Heat rises. The building breathes.

We have followed the token backward through the inference engine, through the silicon, into the building where mathematics becomes physical. The data center is where abstract patterns take form in racks of servers drawing power, generating heat, demanding constant attention from the systems keeping them running.

But buildings are passive. They house equipment; they do not power it. The gigawatt appetite described at the start of this chapter must be fed continuously, around the clock, without interruption. That power must travel through transmission lines that took decades to build. It must originate in turbines and reactors that cannot simply be conjured into existence.

The student's question has traced a path from screen to chip to server to building. Now we follow the copper deeper—through the transformers that step down voltage, through the switchgear that routes power, through the transmission lines that stretch across the Michigan countryside. The next constraint is the grid itself.

The Power Constraint

S ARAH ARRIVED AT PJM Interconnection headquarters on Monroe
Boulevard in Audubon, Pennsylvania, at 5:47 AM on a Tuesday in
December 2024. The parking lot was nearly empty, just a handful of cars
belonging to the overnight operations staff who kept the grid running
while the rest of the region slept. The building—a low-slung corporate
campus surrounded by bare trees and frozen grass—gave no hint of what
happened inside. No visible wires, no humming transformers, no indication that this was the nerve center for the largest power grid in the United
States.

PJM is the regional transmission organization that coordinates the
wholesale electricity grid across thirteen states and Washington, D.C.,
managing power flow for sixty-five million people. As a senior grid operations analyst, *Sarah* had learned to start early. The interconnection
queue waited for no one. This formal process, through which new power
plants and large electrical loads get permission to connect to the grid,
kept piling up requests faster than her team could process them.

She badged through the security turnstile, nodded at the guard who
knew her by name after eleven years, and walked the familiar corridor to
her cubicle. The fluorescent lights hummed at a frequency she no longer
consciously heard. The carpet was the gray-blue of institutions everywhere. A framed photo of her daughter's soccer team sat next to her

monitor; her daughter was fourteen now, the photo three years out of date, but *Sarah* had never gotten around to replacing it.

Her screen loaded the dashboard she had designed three years earlier, when data center requests were still manageable. The numbers glowing there now would have baffled her younger self. The column labeled "Data Center Pipeline" showed 40 gigawatts of power demand under contract or in active negotiation across PJM's thirteen-state territory.[68] Forty gigawatts. Eight times the total electricity consumption of Washington, D.C. The output of forty nuclear reactors running at full capacity. Eighty large natural gas plants. More than half of all U.S. solar capacity.

Sarah took a sip of coffee that had already gone cold—she had bought it at the Wawa on her way in, and it tasted like it always did, adequate and forgettable—and scrolled through the morning's new applications. Three more gigawatt-scale requests had arrived overnight—one from Michigan, two from Virginia. Each would require months of study, represented billions of dollars waiting on her analysis, would transform some community she would never visit.

Eleven years on this job. Her lower back ached from too many hours in this chair. Her eyes needed reading glasses now, though she kept forgetting to wear them. She understood grid topology the way a surgeon understands anatomy. But nothing in her training had prepared her for this. PJM had spent a century building infrastructure to power the eastern United States. Now a handful of technology companies wanted to double that demand in less than a decade.

She clicked on the Michigan request. Related Digital. Saline Township. Something called Stargate. The number in the capacity field made her blink: 1.4 gigawatts.

Depending on how you ran the numbers, that could be close to a quarter of DTE's peak load. From just one facility.

Sarah whistled softly to no one in particular. The sound disappeared into the acoustic ceiling tiles. Welcome to another Tuesday.

She scrolled through the application details, her reading glasses finally on, the text sharp now. The supporting documents included letters from the township, public comment summaries, consent agreement terms. One name appeared repeatedly in the file notes: a township supervisor who had initially voted against the project, then reversed after a lawsuit. She wondered what that vote had cost him. In over a decade at PJM, she had processed thousands of interconnection requests, each one a stack of engineering specifications and legal documents. Somewhere behind every megawatt was a community that would live with the consequences. She had learned not to think about that too much. The queue was long, and the applications kept arriving. Outside her window, the sun was finally rising, painting the frozen parking lot in shades of gray and gold.

Every constraint shapes what can be built. In software, the constraint is often time—engineers race to ship before competitors. In chip manufacturing, physics—how small can you make a transistor? For AI infrastructure, the constraint that dominates all others is power.

Not computing power. Electrical power. The raw flow of electrons measured in megawatts and gigawatts. This constraint determines nearly everything else: where data centers can be built, how quickly they can come online, who can afford to build them, and ultimately how fast artificial intelligence can advance.

Understanding the power constraint requires abandoning some intuitions about how technology companies operate. Google and Microsoft are not primarily limited by how many GPUs they can buy, though chip supply matters. Not by how much capital they can raise, though the sums are staggering. They are limited by how many electrons they can reliably obtain, year after year, at prices that make their business models work.

This chapter explains why power constrains AI progress more than chip supply, more than capital, more than talent. It traces the math behind gigawatt-scale projects, examines why these facilities cannot locate in cities, and maps the byzantine processes through which large loads connect to the grid. By the end, the answer to a seemingly simple question—why build a $7 billion data center on Michigan farmland?—will be clear.

4.1 THE MATHEMATICS OF SCALE

The numbers resist comprehension. America's AI data center buildout encompasses over 600 projects with a combined power capacity exceeding 131 gigawatts. Total announced investment surpasses $1.1 trillion.[8] Nothing in American industrial history compares. Rural electrification in the early twentieth century comes closest, but that transformation unfolded over four decades. This one attempts to compress equivalent change into four years.

Sarah remembered her first week at PJM, more than a decade ago, when a senior engineer named Doug had explained the basics over bad coffee in the break room. "Forget the physics," he had said. "Just remember the scale. A kilowatt runs your toaster. A megawatt runs a small factory. A gigawatt—that's a whole power plant, running flat out, just to keep one customer happy." He had drawn it on a napkin: three zeros between each prefix, each step up meaning a thousand times more infrastructure, a thousand times more planning, a thousand times more ways for things to go wrong.

A primer on electrical power: Electricity is the flow of electrons through wires. Voltage is the pressure pushing them forward; current is how many flow per second. Power—measured in watts—combines both into a rate of energy transfer. Think of it like water in pipes: voltage is the pressure, current is the flow rate, and wattage is how much work the water can do. The units scale by thousands: a kilowatt is a thousand watts, a megawatt is a thousand kilowatts, a gigawatt is a thousand megawatts.

Doug had retired three years ago to a cabin in the Poconos, saying the job had gotten too strange. *Sarah* understood now. When she had started, a fifty-megawatt request was unusual. Now she was staring at 1.4 gigawatts—twenty-eight times larger—from a single applicant.

To make sense of these numbers, we need reference points. A gigawatt of electrical capacity equals a typical nuclear plant running at full output, or San Francisco's peak demand. When a single data center project plans to consume 1.4 gigawatts, it will draw as much electricity as a major metropolitan area.

The Saline Township project in Michigan plans for 1.4 GW of capacity. This represents somewhere between a sixth and a quarter of DTE Energy's current peak load—a substantial share of all the electricity the utility provides to its 2.2 million customers across southeastern Michigan. One facility consuming that much of a major utility's capacity.

Saline Township is not exceptional. It is typical of the new scale. Project Jupiter in New Mexico plans for over 3 GW. Homer City Energy Campus in Pennsylvania targets 4.5 GW. Eleven announced projects each exceed 1 GW of planned capacity, and dozens more approach that threshold.

Why so large? The answer lies in how AI compute scales. Training a large language model requires thousands of high-powered processors working in concert for months. The chips described in Chapter 2—each consuming hundreds of watts—must be packed by the tens of thousands. A modern AI training cluster might contain 25,000 such chips, drawing 17.5 megawatts in GPU power alone. Add cooling, networking, storage, and overhead, and a single training cluster requires 25 to 35 MW of total facility power.

Training is only half the picture. Once trained, these models must serve billions of inference requests—the questions and prompts that users submit. Inference at scale requires even more infrastructure than training. Each request consumes less power than training, but the volume is staggering. A popular AI service might handle billions of requests per day, each requiring milliseconds of attention from specialized hardware.

The economics of AI infrastructure favor concentration. One facility with 100,000 GPUs runs more efficiently than ten facilities with 10,000 GPUs each. Network latency between chips matters for training; keeping chips physically close reduces communication overhead. Cooling systems achieve economies of scale. Security and staffing costs spread across larger footprints. When OpenAI or Google calculates the optimal size for a new AI campus, the math points toward enormous scale.

These calculations produce the gigawatt facilities now proliferating across America. A 1 GW data center can house 400,000 to 500,000 high-power GPU equivalents at current densities, though next-generation chips will consume more power per unit. Enough computing capacity to train multiple frontier AI models simultaneously while serving billions of inference requests daily.

––––––

A critical distinction separates two related concepts: capacity and consumption. Capacity is peak instantaneous demand—how many megawatts or gigawatts a facility can draw at any moment. Consumption is total energy over time: multiply capacity by hours of operation. A one-gigawatt facility running continuously for a year would consume about 8,760 gigawatt-hours.

Data centers care intensely about both, but capacity usually determines location viability. The grid must deliver that full gigawatt-plus to Saline Township at 3 PM on the hottest day of August, when air conditioners across Michigan run at maximum. That average demand might be 30 percent lower matters less than the peak requirement.

Utilities plan for peak demand because electrical systems must balance instantaneously. Unlike other commodities, electricity cannot be economically stored at grid scale—though battery technology is slowly changing this. Every watt consumed must be generated at that exact moment. When a data center adds a gigawatt of peak load to a utility's service territory, the utility must ensure generation capacity exists to meet that load plus a safety margin.

This peak-demand reality explains why utilities speak in capacity rather than consumption. DTE Energy's concern is not primarily how many gigawatt-hours the facility will consume annually—though at full operation, that figure might reach 12 TWh, approximately 12 percent of Michigan's total electricity consumption. The immediate concern is whether DTE can deliver the contracted capacity when needed without compromising service to existing customers.

4.2 WHY NOT CITIES?

If you have followed technology news over the past decade, you might picture data centers as sleek facilities in urban or suburban settings. Companies like Equinix and Digital Realty built their businesses on colocation facilities near major internet exchange points, often in metropolitan areas. These facilities served the cloud era, housing servers for thousands of companies in buildings of 50 to 100 MW capacity.

The AI era has rendered this model obsolete for frontier computing. A 50 MW facility that seemed large in 2015 is now a rounding error. The new facilities require 500 MW, 1,000 MW, 3,000 MW. Loads of this magnitude cannot locate in cities.

City electrical infrastructure cannot deliver gigawatt loads. Urban distribution networks—the wires under streets, transformers on poles—serve thousands of customers at modest power levels. A typical urban substation handles 100 to 200 MW for an entire neighborhood. When a data center needs 500 MW or more, it must bypass the distribution network entirely and connect directly to high-voltage transmission lines. Chapter 5 explores grid topology in detail; what matters here is the consequence. Gigawatt data centers must locate where transmission lines run, not where cities are.

Upgrading urban infrastructure is not viable. Tearing up streets, negotiating easements, relocating existing infrastructure, managing construction in dense areas. Billions of dollars. Five to ten years. Data center developers cannot wait.

———

The alternative is to connect directly to the transmission network, bypassing distribution entirely. Gigawatt data centers locate where high-voltage transmission lines run, build their own substations, and take power from the electrical highway rather than the local roads.

Transmission infrastructure follows predictable geographic patterns. Lines connect power plants to load centers, running through corridors that follow highways, rail lines, or river valleys. These corridors traverse rural areas, where land is cheap and rights-of-way easier to acquire. The irony of AI infrastructure: the most advanced computing facilities on Earth must locate where transmission lines cross farmland, not where technology workers live.

Northern Virginia became the world's first data center megacluster through historical accident. The region happened to have strong transmission infrastructure connecting East Coast power plants to federal facilities around Washington. When early internet companies needed reliable power for their servers, they found it in Loudoun County, where substations and transmission lines existed to serve government installations that never fully materialized. By the time the industry recognized the pattern, Northern Virginia had more data center capacity than the next ten markets combined.

Even Northern Virginia has limits. In late 2024, Dominion Energy reported that eastern Loudoun County—the heart of Data Center Alley—had hit grid saturation. After decades of growth, the region had consumed its available power. New projects now locate in adjacent rural counties, wherever transmission capacity remains.

————

This constraint explains Saline Township. Why would one of the most sophisticated computing facilities ever built locate in a rural Michigan township of 2,300 people? Not near Ann Arbor, with its university. Not near Detroit, with its river and labor force. Power.

DTE Energy's transmission infrastructure crosses Washtenaw County in patterns determined decades ago by industrial geography. The corridor serving Saline Township connects to substantial generation assets and can deliver power at transmission voltages. When Related Digital and Oracle evaluated sites, they did not ask which location had the best highways or the nearest engineering schools. They asked where they could get deliverable power at the scale they needed.

Harold's grandfather cursed those transmission towers when DTE strung them across the back forty in 1957. Ugly steel lattice cutting through his fields, humming in the summer heat, casting shadows he swore confused the soybeans. He accepted the easement payment—a few hundred dollars a year—because the alternative was fighting the utility in court. Three generations of his family farmed around those towers, never quite forgiving them for breaking up the horizon.

Now those towers are why *Harold*'s land commanded a premium. The infrastructure his grandfather resented became his lottery ticket—and like most lottery winners, he would discover that sudden wealth comes with its own burdens. When the developers explained why they wanted his particular five hundred acres, they pointed to maps showing transmission capacity and substation locations. *Harold* understood then that the choice had never been about his farm at all. It was about the electrons that had hummed above his soybeans for sixty years.

The farmland itself is almost irrelevant economically. The site comprises roughly 575 acres, and the entire land purchase totaled perhaps $7 million—about one-tenth of one percent of the project cost. Chapter 7 examines in detail why data centers choose farmland over abandoned industrial sites. Here, the essential point is simpler: if the transmission corridor had crossed a cornfield twenty miles east, the data center would have located there instead. Power determines location. Everything else is secondary.

4.3 THE INTERCONNECTION QUEUE

Locating near transmission infrastructure is necessary but not sufficient. Connecting to the grid requires navigating the interconnection queue— a regulatory and engineering process that has become the binding constraint on new development. Chapter 5 details how this works; here we focus on why it matters for site selection.

The queues have exploded—and data centers face delays on both sides of the meter. On the generation side, PJM's queue contained 265 GW of proposed power plants in late 2024—solar farms, wind projects, gas turbines waiting to connect. The total generation queue across major grid operators exceeded 2,000 GW, roughly twice the entire generating capacity of the United States.[69] Most of these projects will never complete—historically, only about 10 to 15 percent have successfully connected to the grid. FERC issued Order 2023 to accelerate generator interconnection, but results remain mixed.

On the load side, utilities face their own surge of large customer requests. ERCOT's large load queue—dominated by data centers—grew from 63 GW in late 2024 to over 200 GW by late 2025.[70] Wait times for both generation and load connections have stretched from two to three years in 2008 to five to eight years today. Some projects face ten-year waits. For a technology industry accustomed to six-month product cycles, this pace is incomprehensible.

Texas offers a partial counterexample. ERCOT operates independently from the eastern and western interconnections, providing regulatory flexibility despite vulnerabilities exposed by the 2021 winter storm. ERCOT's interconnection process moves faster, with some projects completing in eighteen to twenty-four months. Texas has also shown greater tolerance for behind-the-meter generation, where data centers build their own power plants on-site. Projects that might wait seven years in Virginia can begin operations in two years in Texas—a key factor in the state's 10.9 GW pipeline. Chapter 5 explores the regional transmission organizations (RTOs) and their rules in detail.

Interconnection is not only slow but expensive. Projects must pay for grid studies, typically hundreds of thousands to millions of dollars. If studies reveal needed infrastructure upgrades, the project must contribute to those costs—which can reach hundreds of millions for transmission line construction or substation expansion.

PJM allocated $4.3 billion in infrastructure costs to customers in 2024, much driven by data center connections.[71] Virginia alone accounted for $1.98 billion—46 percent of the regional total. These costs exclude the data centers' own investments in substations and on-site infrastructure.

The magnitude of these costs helps explain why smaller projects struggle to compete. A startup with $100 million in funding cannot absorb $50 million in interconnection costs plus years of delay. Only the largest, most patient capital can navigate the process. This filtering effect concentrates AI infrastructure among a handful of well-funded players.

4.4 POWER FIRST

Within the data center industry, a principle has emerged from hard experience: power first. Available electrical capacity is the primary site selection criterion, outranking all other factors by a wide margin.

Industry surveys confirm this shift.[72,73] Grid availability and infrastructure delivery timelines now outweigh traditional factors like fiber connectivity, workforce availability, and tax incentives. Power dominates every other consideration.

––––––

When a hyperscaler like Microsoft or Google plans a new AI campus, the process begins with power, not real estate. Teams first identify utility territories with available capacity, examining integrated resource plans, queue positions, and transmission topology. They assess which utilities are cooperative, which face regulatory constraints, and which have realistic timelines for new connections.

Only after identifying promising utility territories do teams evaluate specific sites. Within a qualifying territory, they seek land near transmission corridors, preferably with existing substations that can be expanded. Highway access, fiber availability, and labor markets matter—but they cannot overcome power deficits.

This approach produces locations that would seem bizarre by traditional corporate real estate logic. Major technology companies are building their most valuable facilities in Junction City, Kansas; Papillion, Nebraska; and Saline Township, Michigan—communities that never appeared on any corporate relocation shortlist. These locations won not because of their airports or talent pools but because electrons flow through their territories.

––––––

The power-first approach elevates utilities to a role they have never occupied. In traditional site selection, utilities were service providers, taken for granted. A factory might locate in a state for tax reasons and expect the local utility to provide power as a matter of course.

For gigawatt data centers, utilities are partners. The relationship begins years before construction and extends decades into the future. Developers cultivate utility executives, negotiate complex contracts, and structure deals requiring regulatory approval. A utility's willingness and ability to serve large loads becomes a competitive differentiator among regions.

Dominion Energy has positioned itself as the premier utility partner for hyperscale data centers, parlaying its Northern Virginia position into a 40 GW development pipeline.[74] The company has raised its five-year capital expenditure plan to $50.1 billion to serve data center growth, with $41 billion allocated to Virginia operations. This level of investment transforms Dominion from a traditional utility into an active participant in the AI infrastructure buildout.

Other utilities are following Dominion's lead. Georgia Power announced a $16 billion grid expansion in December 2025, explicitly designed to enable data center development.[75] Entergy invested $4.5 billion in Louisiana infrastructure tied to Meta's Hyperion campus.[76] AEP, Duke, and dozens of other utilities have identified data centers as growth engines worth billions in infrastructure investment. But utility enthusiasm has a flip side: ratepayer concern about who bears the costs. We return to this tension below.

Capital has always needed physical outlets. When railroads ran out of western frontier, money flowed into urban real estate. When urban markets saturated, capital opened the suburbs through highways and mortgages. When suburban sprawl hit limits, money poured into global supply chains and emerging markets. Each generation of American capitalism finds its geographic fix—new terrain where accumulated wealth can be invested, where construction creates profit, where infrastructure opens new possibilities for accumulation.

Now data centers open transmission corridors. The geographer David Harvey calls this pattern a "spatial fix"—capital's structural need to expand into new geographies when profitable investments grow scarce elsewhere.[77] The AI boom has generated trillions in market value, but those trillions need physical expression. Server halls. Cooling systems. Transmission lines. Substations. The infrastructure must go somewhere, and "somewhere" turns out to be farmland near transmission corridors—the next frontier for capital seeking returns.

The 40 gigawatts that *Sarah* monitors each morning at PJM represent spatial fix in motion. Each application in her queue is capital searching for a place to land—money that accumulated in tech stocks and venture funds now seeking physical form in concrete and copper and cooling water. The pattern would be familiar to anyone who watched the railroad boom of the 1870s or the highway boom of the 1950s. Only the technology is new. The underlying logic—capital's need to keep moving, keep building, keep finding new terrain—is as old as capitalism itself.

4.5 WHY POWER TRUMPS CHIPS

A natural question emerges: is power really the binding constraint? Chip supply matters. NVIDIA's GPUs are famously difficult to obtain, with waitlists extending months and prices elevated by scarcity. Could chips, rather than power, be the true limit on AI progress?

The evidence suggests otherwise, for several reasons.

First, chip supply is expanding. NVIDIA shipped approximately 3.8 million data center GPUs in 2023, growing to roughly 4 million in 2024, with production continuing to expand as new fabrication facilities come online.[78] AMD, Google, and Intel have all expanded AI accelerator production. More important, chips can be stockpiled—companies can place orders years in advance. These chips lose value while sitting idle, but power infrastructure takes even longer to build. In mid-2024, Microsoft CEO Satya Nadella acknowledged that the company had "inventory that we can't even light up"—chips sitting idle because the data centers to house them were not ready.[79]

Second, power cannot be stockpiled. Electricity cannot be stored at scale. Every megawatt consumed must be generated and delivered at that exact moment. A data center filled with GPUs but lacking grid connection is stranded capital. Power is a flow constraint, not a stock constraint.

Third, power timelines exceed chip timelines. Chip supply constraints typically operate on one-to-two-year horizons. Power infrastructure operates on much longer cycles: three to seven years for new generation, five to ten years for major transmission lines, three to eight years for interconnection studies. No amount of money makes electrons arrive faster. By the time a data center completes its power infrastructure, chip supply will likely have improved. A company with abundant chips and no power has nothing.

David understood this before most of his peers. His eureka moment came in early 2024, reviewing a deal that had fallen through. The project had everything: anchor tenant commitment, GPU allocation locked, construction team ready. What it lacked was grid access. The interconnection queue showed a seven-year wait. The deal died.

He started calling utility executives. The conversations were revealing. One, at a Midwestern utility, explained the mathematics bluntly: "You cannot build transmission capacity faster than we can study what you need. The physics do not care about your investment timeline." Another described the queue backlog—hundreds of gigawatts waiting, years of processing ahead.

"That was when it clicked," *David* would later explain. "Chips depreciate. Land appreciates. But power access? Power access is the moat. A project with secured grid capacity is worth more than one with GPU allocations, because chips you can eventually buy. Power takes years." His fund pivoted to prioritizing projects with interconnection agreements already in place or well advanced in the queue. The strategy would prove prescient.

Capital allocation reveals where constraints bind. When a resource is the limiting factor, companies invest heavily to secure it. The investment pattern in AI infrastructure clearly identifies power as the constraint.

Hyperscaler capital expenditure has reached record levels, with the largest companies each committing tens of billions annually. Chapter 8 provides the full breakdown. These budgets dwarf expenditures on chips themselves. The majority targets power infrastructure: substations, transmission lines, cooling systems, grid connections.

Hyperscalers are also investing in power generation directly, pursuing nuclear partnerships that would have seemed fanciful a decade ago. Amazon, Microsoft, and Google have collectively committed billions to restart dormant plants and develop next-generation reactors—investments detailed in Chapter 6. These commitments make sense only if power, not chips, is the scarcer resource.

4.6 POWER FIRST IN ACTION

The power-first principle manifests throughout the AI infrastructure buildout. Three examples illustrate the pattern.

Northern Virginia hosts more data center capacity than any other market on Earth—over 250 facilities with approximately 4 GW of connected load. The region's dominance stems from historical infrastructure advantages that compounded over decades. But even Data Center Alley has limits: eastern Loudoun County has hit transmission saturation, forcing new projects to adjacent counties where capacity remains. Chapter 5 details how grid topology shapes these geographic patterns.

Texas presents a different variation. ERCOT's faster interconnection and abundant renewable resources have attracted substantial development—the state's 10.9 GW pipeline makes it the second-largest market after Virginia. But Texas projects cluster in specific locations determined by grid topology. Dallas-Fort Worth hosts the majority of hyperscale facilities; West Texas, despite abundant land, remains challenging due to limited transmission.

The market-driven ERCOT model has also produced a distinctive response: behind-the-meter generation. Over half of the gigawatt-scale projects announced in 2024 and 2025 include on-site power, usually natural gas turbines.[8] These projects bypass the interconnection queue by generating their own power—solving the timeline problem at the cost of added carbon emissions.

The Saline Township project demonstrates how power can redirect investment to unexpected locations. Michigan has never been a major data center market. Its utility infrastructure developed to serve manufacturing, not computing. Yet when the Stargate initiative evaluated sites, power availability dominated. DTE Energy's transmission corridor through Washtenaw County offered access to substantial capacity. The utility expressed willingness to invest hundreds of millions in infrastructure upgrades, financed by the project itself.

Michigan also offered policy support. Senate Bill 237, signed in late 2024, exempts qualifying data center investments from sales and use taxes through 2050.[80] This incentive, valued at approximately $90 million for the Saline project, improved economics but did not determine location.[81] Power determined location. Incentives improved returns. The $7 billion investment represents the largest single private project in Michigan history.

4.7 THE RATEPAYER QUESTION

The power constraint connects data centers to communities of existing customers. The costs of serving gigawatt loads ripple through utility rate structures, raising questions about who benefits and who pays.

When a utility builds infrastructure to serve a data center, those costs enter the rate base—the total capital investment on which the utility earns a regulated return. Should existing residential and commercial customers subsidize infrastructure built primarily for data centers? Utilities argue that data centers provide system benefits. Critics counter that socializing costs creates an unfair burden on households and small businesses.

PJM's 2024 capacity market auction highlighted the stakes. Total capacity market costs rose from $2.2 billion to $14.7 billion—more than a sixfold increase driven substantially by data center demand.[71] These costs flow through to all PJM customers, including residential ratepayers.

Virginia has become the testing ground for cost allocation policy. The state legislature has demanded that Dominion Energy create special rate classes isolating data center costs from residential customers. The regulatory outcome will set precedent nationwide.

Michigan's MPSC approval of the Saline Township power contracts illustrated these tensions. DTE Energy claimed the project would provide a $300 million net benefit to other customers, arguing that data center revenue would reduce per-customer infrastructure costs and improve system economics.

Attorney General Dana Nessel disputed this claim, noting that contract details were heavily redacted and verification impossible.[7] Community advocates argued that 5,500 public comments opposing the project were ignored in favor of rapid approval.[82] The ex parte process—approval without contested hearings—drew criticism as regulatory capture favoring corporate interests over transparency.

The fundamental dispute concerns risk allocation. DTE claims that if the data center fails to pay its obligations, costs cannot be passed to other customers. But the contract details supporting this claim remain confidential. If DTE's cost estimates prove wrong, or if the data center underperforms, who bears the loss? The opacity of these contracts leaves ratepayers uncertain about their exposure.

The examples above—Virginia, Texas, Michigan—illustrate a broader pattern. Power availability predicts where data centers locate with remarkable precision. A map of announced projects shows dense clusters in Northern Virginia, the Texas Triangle, and greater Phoenix. Regions with data center clusters attract utility investment that attracts additional development. Regions without clusters face a chicken-and-egg problem.

Looking forward, power constraints will intensify before they ease. Announced projects far exceed current grid capacity. Even among publicly announced projects—a more vetted subset than raw queue entries—industry experience suggests only 30 to 40 percent reach operation. The gap between announced capacity and deliverable power creates market dynamics favoring early movers. Projects that secure grid access early gain competitive advantages that compound over time. The hyperscalers' asymmetric power—described in Chapter 3—stems partly from this dynamic: they started building relationships with utilities years before their competitors recognized the constraint.

4.8 Power Determines Everything

We began with a question: why build a $7 billion data center on Michigan farmland? The answer, now clear, is power.

The Saline Township site won Stargate's nationwide competition not because of Michigan's universities, workforce, or tax incentives—though all mattered at the margin. It won because DTE Energy's transmission infrastructure crosses that farmland with capacity to serve 1.4 gigawatts of load. The electrons flow there. Everything else follows.

This insight—power determines everything—opens the rest of the AI infrastructure story. Why have data centers located on farmland rather than in cities? Because transmission corridors cross farmland. Why have projects taken five to ten years from announcement to operation? Because interconnection and construction have required that time regardless of capital deployed. Why have wealthy technology companies partnered with regulated utilities? Because electrons flow through utility wires. There has been no alternative.

Understanding the power constraint clarifies the economic geography of AI development. It explains why certain regions attract billions while others are bypassed, why state incentives matter less than utility cooperation, why the environmental impacts of AI extend far beyond the data centers themselves to the generation sources that power them.

The next chapter examines the grid—the system that moves power from generation to consumption.

CHAPTER FIVE

The Grid

T HE SUMMER THUNDERSTORM ROLLED ACROSS Northern Virginia at 4:47 PM on a Wednesday in July 2025, and *Sarah* watched the transmission map on her screen turn yellow, then orange. Eleven years at PJM Interconnection, working from the control room in Valley Forge where wall-sized displays tracked electricity flowing across thirteen states. She had seen heat waves. She had seen polar vortexes. She had never seen anything like the past two years.

Loudoun County was spiking. Data Center Alley pulling hard.

Sarah pulled up the detail view. Six gigawatts of operational load concentrated in a single county, with another six gigawatts under construction—Dominion had contracted for 40 gigawatts across Virginia, a pipeline that would eventually dwarf most states' total consumption. A decade ago, that corridor drew perhaps two gigawatts. The AI boom had tripled demand, and now a thunderstorm was threatening the transmission lines feeding that load.

Lightning struck near the Goose Creek substation. A 500-kilovolt line tripped. For a fraction of a second, fifty million dollars worth of computing equipment teetered between operation and shutdown. Then the automated systems responded: backup circuits engaged, data centers switched to battery power, and diesel generators across the region coughed to life.

Sarah exhaled. The internet stayed up. Somewhere, millions of AI queries continued without interruption, their users never knowing how close they had come to seeing spinning wait cursors.

"Non-event," she typed into the incident log. But she knew better. The margins were shrinking. Every new data center that came online, every gigawatt of AI demand added to the grid, reduced the buffer between normal operation and catastrophe. The infrastructure that kept the digital economy running had been built for a different era. Understanding that infrastructure—how it works, why it cannot be easily expanded, and what happens when it hits its limits—is essential to understanding why data centers are built where they are built.

5.1 THE GRID IS NOT ONE THING

Americans speak of "the grid" as if it were a single, unified system. It is not. What we call "the grid" is actually three weakly connected grids. Dozens of entities operate them. Federal and state jurisdictions overlap in governing them. And the whole thing was built over a century by thousands of local decisions that had nothing to do with twenty-first-century computing.

The three major grids—the Eastern Interconnection, the Western Interconnection, and the Texas Interconnection (known as ERCOT)—share a common electrical frequency of sixty hertz but have very limited transfer capacity between them. A power plant in California cannot easily send electricity to Chicago. A generator in Texas cannot help when New York faces a heatwave. For practical purposes, each interconnection is its own continent.

The Eastern Interconnection is by far the largest, serving over 200 million Americans and stretching from the Atlantic seaboard to the Rocky Mountain foothills, from Canada to Florida. Within this vast territory, dozens of utilities and grid operators coordinate the second-by-second balance of electricity supply and demand. The Western Interconnection serves the Pacific Coast states and the Mountain West. ERCOT serves Texas—and only Texas, by deliberate design.

Within each interconnection, utilities control service territories— owning the power plants, operating the transmission lines, selling electricity to customers. These utilities range from massive investor-owned corporations like Dominion Energy and DTE Energy to tiny rural electric cooperatives serving a few thousand farms. The Tennessee Valley Authority, a federal corporation created during the New Deal, serves parts of seven southern states. Municipal utilities serve Los Angeles, Sacramento, and Austin.

This patchwork reflects American federalism and American history. Electricity regulation splits between the Federal Energy Regulatory Commission (FERC), which governs wholesale markets and interstate transmission, and state public utility commissions, which regulate retail rates and approve new generation. Building any significant piece of grid infrastructure means navigating multiple regulatory bodies. Different priorities. Different timelines. Different constituencies.

A transmission line crossing state boundaries might need approval from two state utility commissions, FERC, multiple county governments, and various environmental agencies. Each process operates on its own timeline, applies its own standards, and creates its own opportunities for opposition. A determined opponent can delay a project for years by finding the weakest link in this chain.

For data center developers, this fragmentation creates both opportunities and obstacles. Power availability varies enormously from one location to another—a site with abundant capacity in Virginia might face a five-year wait for interconnection in Michigan. State incentive programs can matter as much as grid capacity. And understanding the specific regional operator matters more than understanding "the grid" in general.

The grid we have today was not designed. It grew—organically, chaotically—over more than a century, with no master plan guiding its development. Utilities built incrementally to serve local customers, connecting their systems to share backup power. Blackouts forced coordination: the Northeast blackout of 1965 led to regional reliability councils, and the North American Electric Reliability Corporation established common standards. But these remain coordination mechanisms, not control mechanisms. No single entity operates the American grid. The infrastructure was built for gradual residential load growth, not for single customers appearing overnight demanding 500 megawatts of continuous power.

5.2 REGIONAL OPERATORS

Over the past three decades, much of the grid has been reorganized under RTOs and independent system operators (ISOs)—different names for essentially the same function. These entities own no power plants, no transmission lines. Instead, they operate the wholesale electricity markets where generators sell power, direct the flow of electricity across the grid in real time, and manage the process of connecting new loads and new generators to the system.

RTOs and ISOs matter enormously for data centers because they control two things: the price of electricity and the timeline for interconnection. Understanding these regional operators explains why data centers cluster where they do.

———

PJM Interconnection is the largest RTO in North America, serving 67 million people across thirteen states and the District of Columbia—from New Jersey to Illinois, from Michigan to North Carolina. The name comes from its origins: Pennsylvania, Jersey, Maryland. Today, PJM coordinates a $50 billion wholesale electricity market and manages 88,000 miles of transmission lines.[83]

More than half of all American data center capacity sits within PJM's territory. This is no accident. Northern Virginia's Data Center Alley developed there because of decisions made decades ago. Where to route fiber. Where to build internet exchange points. Which utilities would serve unusual customers needing enormous amounts of very reliable power. Once the first data centers arrived, network effects took over. Each new facility made the location more attractive for the next.

That concentration has created the crisis Chapter 4 described. The interconnection queue—the list of generators and large loads waiting to connect—has swelled to more than double the system's current peak demand. Wait times have stretched from two years in 2008 to eight years or more today. The queue is so clogged that PJM adopted emergency reforms in 2024, abandoning first-come, first-served for a clustered study approach that evaluates groups of projects together.

Inside PJM's control room in Valley Forge, wall-sized screens display electricity flows across the entire territory—demand rising and falling in real time, generators ramping up and down, power flowing across transmission lines that operate near their limits.

Sarah remembers when data centers were a rounding error in demand forecasts. She started here as a summer intern during engineering school at Penn State, back when operators worried about steel mills and auto plants. When a factory closed, demand dropped. When a new mall opened, demand ticked up. Predictable. Gradual.

Now she gestures at a cluster of red dots concentrated in Northern Virginia. Six gigawatts of load in one county, with six more under construction and contracts for 40 gigawatts across Virginia. PJM has had to add a full-time analyst just to track what is being built out there. The morning forecast shows data center load growing faster than any other category. The infrastructure was designed for 1 percent annual growth, not 15 percent. The queue is ridiculous—companies file applications knowing they will wait five, six, seven years. Some file at multiple sites just to hold their place. The utilities serving PJM territory cannot build fast enough.

For data center developers, PJM's queue backlog means that the traditional approach—buy land, apply for interconnection, wait for the utility to extend service—no longer works. The timeline is simply too long. This has driven the industry toward alternatives: behind-the-meter generation that bypasses the queue entirely, or locations in less congested grid regions where interconnection moves faster.

———

The Electric Reliability Council of Texas operates the Texas grid as an island, deliberately disconnected from the rest of the country. This isolation exists because of a quirk of regulatory history: by staying within state boundaries, ERCOT avoids federal jurisdiction over its wholesale market. Texas utilities preferred state regulation to federal oversight, and they built their grid accordingly.

The Texas grid serves 26 million people and coordinates roughly 90 percent of the state's electricity load. It operates as a single market where generators compete to sell power and consumers benefit from some of the lowest electricity prices in America. The regulatory philosophy emphasizes market competition over utility planning: generators build power plants when they expect to profit, not when regulators order them to.

For data centers, ERCOT's isolation creates both advantages and risks. The advantage is speed: ERCOT's interconnection process moves faster than other regions, often completing in months rather than years. Texas also offers abundant land, favorable tax treatment, and a business-friendly regulatory environment. Companies can build quickly and operate with minimal oversight.

The risk is reliability. ERCOT's isolation means it cannot import power from neighboring grids during emergencies. When Winter Storm Uri struck in February 2021, the Texas grid came within minutes of total collapse. Rolling blackouts left millions without power for days in subfreezing temperatures. At least 246 people died according to official state counts, though researchers estimate the true toll ran far higher.[84,85] Natural gas wells froze. Wind turbines iced over. Power plants tripped offline in a cascade that nearly became catastrophic.

The fundamental vulnerability was not weather but design. ERCOT's market structure pays generators only for the power they actually sell, not for maintaining reserve capacity. This creates incentives to build just enough capacity for normal conditions, without margins for extreme events. When extreme weather hit, those margins did not exist.

Since Uri, ERCOT has added generating capacity and tightened winterization requirements. Generators must now weatherize equipment and face penalties for failures during emergencies. Reserve margins have improved. But the fundamental vulnerability remains: an isolated grid cannot be rescued by its neighbors. For data centers that promise 99.999 percent uptime, this creates difficult calculations. The speed of Texas interconnection must be weighed against the tail risk of catastrophic grid failure.

Despite these concerns, Texas remains one of the fastest-growing data center markets in America. ERCOT reports inquiries totaling more than 230 gigawatts of large load requests, over 70 percent from data centers— more than twice the state's current generating capacity.[86] Much of this demand is speculative. But even if a fraction materializes, it will transform the Texas grid. Infrastructure built over decades to serve gradually growing residential and commercial demand would need to nearly double to serve the AI boom.

The Midcontinent Independent System Operator (MISO) covers fifteen states from Montana to Louisiana, including Michigan. Like PJM, MISO operates wholesale electricity markets and manages interconnection queues. Its territory spans rich agricultural land, significant wind resources, and industrial centers from Minneapolis to New Orleans. The

242-gigawatt backlog is smaller than PJM's but still represents years of waiting for new projects.[69]

MISO's territory has grown increasingly attractive for data center development. The Midwest offers cheap land, cold winters that reduce cooling costs, and utilities eager for large customers to replace declining industrial loads. Michigan, Ohio, Indiana, and Wisconsin have all seen surging interest. The interconnection process, while slow, is less congested than PJM's. Competition for these projects has intensified as developers look beyond the saturated Virginia market.

The Southwest Power Pool (SPP) covers the Great Plains from North Dakota to Oklahoma. Its territory includes some of the best wind resources in America, making it attractive for data centers seeking renewable energy. SPP has recently created fast-track programs for data center interconnection, recognizing the economic opportunity. The organization has simplified study processes and worked to accommodate large loads more efficiently than traditional approaches allow.

For data center developers, the choice of ISO determines the rules of the game. Each operator has different study processes, different cost allocation methods, different timelines. A project that takes three years in ERCOT might take eight in PJM. The differences are not about technology or geography—they are about regulatory process and institutional capacity. Understanding them is essential to site selection.

5.3 TRANSMISSION VS DISTRIBUTION

The electrical grid operates at two distinct levels: transmission and distribution. Understanding the difference explains why data centers cannot simply be built anywhere with available land and fiber connectivity.

Transmission lines carry electricity at high voltages over long distances. You have seen them: towering steel structures marching across the countryside, cables humming with 115,000 to 765,000 volts. High voltage minimizes energy losses, a physics constraint that determines the grid's architecture. These lines connect power plants to major substations serving cities and industrial areas. They form the backbone, moving power from where it is generated to where it is consumed.

Why high voltage? Electrical losses in a wire are proportional to current squared, but power equals voltage times current. Step up voltage, reduce current, and you transmit the same power with far lower losses. A 500-kilovolt transmission line carries power efficiently for hundreds of miles. A 120-volt household circuit loses significant energy over a few thousand feet.

Distribution lines carry electricity at lower voltages for the final leg to homes and businesses. Wooden poles along residential streets. Cables at 4,000 to 35,000 volts. Distribution systems are designed for small, scattered loads: houses drawing 2 kilowatts on average, commercial buildings drawing tens or hundreds of kilowatts.

Distribution systems reflect their origins. A typical residential transformer serves perhaps a dozen homes, each drawing a few kilowatts. A neighborhood substation might serve a few thousand homes, drawing 5 to 10 megawatts total. The wires, transformers, and switches are sized accordingly—never designed for a single customer drawing 500 megawatts.

Grid operators use a drinking-straw analogy: transmission lines are the main pipes bringing water from the reservoir, distribution lines are the thin pipes within your house. You cannot fill a swimming pool through your kitchen faucet. The pipes are simply too small.

Data centers face the same problem. A modern AI training facility might need 500 megawatts or more—the equivalent of 400,000 homes. Distribution systems cannot deliver that kind of power. The transformers are too small, the wires too thin. The infrastructure was designed for a different era and a different purpose. Upgrading a distribution system to serve data center loads would mean rebuilding it from scratch.

This is why data centers must connect to the transmission system, typically at 138 kilovolts or higher. Such connections require building new substations or expanding existing ones, running new transmission lines from the substation to the site, and engineering studies to ensure the additional load will not destabilize the grid. Each step takes time and money.

The economics are revealing. A new substation serving a 500-megawatt data center costs $50 million to $200 million. New transmission lines run $2 million to $5 million per mile. The interconnection process itself—studies, permits, utility coordination—costs $5 million to $20 million and takes three to seven years. For a company racing to monetize AI compute, these timelines are unacceptable.

5.4 THE QUEUE: WAITING FOR POWER

The interconnection queue is where data center ambitions meet infrastructure reality. Every generator that wants to sell power to the grid, and every large load that wants to buy power, must apply to the relevant ISO or RTO for an interconnection study. The study determines what grid upgrades are needed, who pays for them, and when the facility can begin operating.

In theory, the queue is a rational allocation mechanism. In practice, it has become the bottleneck that shapes the entire industry. Chapter 4 presented the staggering backlog numbers. Here we examine why the queue moves so slowly.

Four factors converged to create the congestion. The energy transition flooded queues with renewable projects, each requiring a full interconnection study, overwhelming ISO staffing. Speculative applications clog the system as developers reserve queue positions at multiple sites to preserve optionality. Data center load growth added enormous continuous demand to already-congested queues, with single facilities drawing as much power as medium-sized cities. And transmission infrastructure has not kept pace—new lines take a decade from conception to completion.

The queue backlog is not merely an inconvenience—it is a binding constraint on the entire AI infrastructure buildout. Companies that need gigawatts of power cannot wait eight years. They must find alternatives or abandon projects entirely.

5.5 THE STUDY PROCESS

The interconnection process involves three stages, each with its own timeline and costs. First comes the feasibility study (three to six months, $10,000 to $50,000), which assesses whether connection is technically possible. Second, a system impact study (six to twelve months, $50,000 to $500,000) examines how the new connection affects the broader transmission network—whether it will overload lines, degrade voltage stability, or reduce reliability. Third, a facilities study (six to twelve months, $100,000 to $500,000) engineers the specific interconnection equipment and produces a binding cost estimate.

Only after completing all three studies can the project sign an interconnection agreement and begin construction. The construction itself takes another one to three years.

The cumulative timeline is daunting: two to four years of studies followed by one to three years of construction, totaling three to seven years from application to power. Study fees, deposits, and charges can reach $20 million to $50 million before any construction begins. Network upgrades might add another $50 million to $200 million. A project might discover halfway through the study process that required upgrades make the site economically unviable. For a company that expected to be operational in eighteen months, this represents a fundamental mismatch between business expectations and infrastructure reality.

5.6 THE FARMLAND PARADOX

Here is the paradox at the center of data center development: despite policy preferences for brownfield redevelopment, despite environmental concerns about consuming productive farmland, data centers overwhelmingly rise on pristine agricultural land.

America has thousands of abandoned industrial sites—former steel mills, coal plants, manufacturing facilities—sitting idle. Many are in communities desperate for economic development. Policy at every level encourages their reuse. Yet data center developers choose cornfields over contaminated brownfields, time after time. Why?

The answer lies in grid topology. Brownfield sites often have worse power infrastructure than greenfield sites. The intuition that "former industrial sites have existing power" is usually wrong. Industrial sites were served by dedicated utility connections that have since been decommissioned. Substations were sized for twentieth-century industrial loads, not twenty-first-century data center demands. When capacity requirements change, queue positions reset. A steel mill that operated at 100 megawatts decades ago cannot serve a data center needing 500 megawatts—the transformers are wrong, the switching equipment is obsolete, the transmission lines are inadequate.

Compare this to farmland near transmission corridors. Environmental assessment is quick and clean. Title is straightforward. The land is flat. And the path to power is faster. Chapter 7 explores the full economics of the farmland paradox—the contamination costs, title complications, and timeline penalties that make brownfield development structurally disadvantaged. What matters here is the grid connection: farmland near transmission infrastructure offers faster interconnection than urban brownfields lacking adequate capacity.

This explains why Saline Township's cornfields attracted Oracle and OpenAI rather than some abandoned factory in Detroit. The township sits near existing DTE transmission infrastructure. The land was clean, the title clear, and the path to power faster than any urban alternative could offer.

5.7 BEHIND THE METER

Faced with multi-year interconnection queues, data center developers have turned to a different approach: generating their own power on-site, behind the utility meter.

A behind-the-meter facility connects to the grid but also has its own power generation—typically natural gas turbines—capable of meeting most or all of its load. The facility might draw from the grid when prices are low and generate its own power when prices are high. Or it might run its own generators continuously and use the grid only for backup.

The appeal is obvious: bypass the queue. Building a gas turbine plant takes two to three years. Connecting to the grid through normal channels takes five to seven years or more. For companies racing to deploy AI infrastructure, the math is compelling—power in 2027 with on-site generation, or power in 2032 from the grid. Chapter 4 documented the result: over half of gigawatt-scale projects now include on-site generation. For its largest projects, the industry has effectively abandoned the traditional utility model.

By late 2025, the scale had grown beyond anything utilities anticipated. Oracle contracted with VoltaGrid for 2.3 gigawatts of modular natural gas generation: ninety-two power packs, each containing reciprocating engines, deployed entirely off-grid.[87] The Joule project in Utah broke ground on 1.3 gigawatts of behind-the-meter capacity, fully islanded from Rocky Mountain Power, with plans to scale to 4 gigawatts.[88] xAI's Colossus facility in Memphis operated dozens of gas turbines for months before receiving permits, drawing criticism from environmental advocates but demonstrating how quickly behind-the-meter power could deploy.[89] The industry had moved from bypassing interconnection queues to building what some analysts call "hypergrids"—gigawatt-scale private power systems capable of operating independently of public utilities.

Natural gas dominates behind-the-meter generation. Of the multi-gigawatt data center projects announced through 2025, nearly all use natural gas as their primary power source. The reasons are straightforward: gas turbines are proven technology, deploy quickly, and provide the reliable baseload power that AI training demands. Solar and wind cannot deliver continuous power without massive storage systems that remain expensive and space-intensive. Nuclear operates on decade-long timelines that exceed the AI industry's patience.

Western Pennsylvania illustrates the pattern. Three projects there—Homer City, TECfusions Keystone, and Shippingport—together plan more than 10 gigawatts of behind-the-meter gas generation, drawing fuel from the Marcellus Shale deposits directly beneath them. Chapter 6 examines these projects in detail; here we note only that they bypass the grid constraints described above by generating their own power.

The environmental implications are significant—Chapter 6 quantifies the carbon footprint in detail. The tension between speed and sustainability defines the current moment. The press releases announce renewable commitments; the permit applications request gas turbines. There is no contradiction—only a gap in timing. The gas plants run now; the renewables arrive later.

The behind-the-meter trend creates new regulatory challenges. Traditional utility regulation assumes utilities build power plants and serve customers through the grid. When customers build their own power plants, the utility loses sales but may still need to maintain transmission capacity for backup power. The economics of the regulated utility model grow strained.

Regulators are beginning to respond. In December 2025, FERC directed PJM to reform its rules governing generation co-located with load, finding that existing behind-the-meter frameworks were "no longer just and reasonable" at data center scale.[90] Texas Senate Bill 6 requires large loads—including behind-the-meter facilities—to pay retail transmission charges on peak demand regardless of self-generation.[91] But these responses remain piecemeal. No standardized framework exists for the behind-the-meter data center era, and the gap between regulatory adaptation and industry practice continues to widen.

5.8 THE SALINE TOWNSHIP CONNECTION

Return to Saline Township. Chapter 4 established the power constraint that drew this project to Michigan farmland. Here we examine the grid mechanics: how does DTE Energy actually plan to deliver the contracted capacity?

DTE operates within the MISO region. MISO's interconnection queue, while less congested than PJM's, still represents years of waiting for large loads. DTE's own data center inquiry pipeline reportedly exceeds 3 gigawatts. Yet the utility claims it can serve the Saline facility from existing resources, augmented by battery storage financed by the project.

The claim strains credulity. Serving a load equal to fifteen to twenty-five percent of DTE's peak capacity without new generation would require either massive efficiency gains elsewhere in the system or substantial unused capacity. Neither explanation fully satisfies. Utilities typically maintain reserve margins of 15 to 20 percent above peak demand. Taking on a single customer of this magnitude implies negligible reserves for normal operations.

More likely, the project benefits from a combination of factors: existing transmission infrastructure, DTE's willingness to prioritize this customer over others, and the project's commitment to finance infrastructure upgrades that might otherwise wait decades. The investment scale and political attention create advantages ordinary customers do not enjoy.

The transmission corridor running through Saline Township is the hidden asset that made the project possible. High-voltage transmission lines cross the township, connecting substations in the broader DTE network. These lines were built decades ago to serve the industrial loads of southeast Michigan. As manufacturing declined, capacity became available—an opportunity that would not exist on an unconnected site.

This is rural Michigan farmland, but rural farmland with high-voltage transmission lines already in place. The grid connection that would take years elsewhere could happen faster here because the infrastructure exists. A data center developer looking at Saline Township sees not just 250 acres of soybeans but a 138-kilovolt transmission corridor with available capacity.

This pattern repeats across the industry. The sites that win projects are not simply those with available land or favorable taxes—they are sites where the grid can accommodate enormous loads without years of upgrade work. Transmission corridors crossing farmland create opportunities urban brownfields cannot match. The map of transmission infrastructure is, in a sense, the map of potential data center locations.

5.9   THE COST QUESTION

Chapter 4 introduced the ratepayer question: who pays for infrastructure that primarily benefits private companies? Here we examine the grid-specific mechanisms in detail.

The cost-shifting flows through capacity markets. Utilities purchase capacity from generators to ensure sufficient supply during peak periods. When demand grows—as it does when data centers arrive—more capacity is needed, and that cost flows through to retail rates. A homeowner in suburban Philadelphia pays higher electricity bills because of data center growth in Virginia.

The numbers are substantial. The Union of Concerned Scientists estimates that utilities in seven PJM states passed more than $4.3 billion in data center-driven transmission connection costs to customers in 2024, with Virginia alone accounting for $1.988 billion.[92] The concentration of data centers in Virginia means Virginia ratepayers bear a disproportionate share of grid upgrade costs.

For individual households, the impact is meaningful. Residential rates in Dominion Energy territory have increased $20 to $40 per month over the past several years, driven partly by data center infrastructure investments. These increases are not itemized on customer bills—they appear simply as higher base rates, with no indication of their cause.

The political backlash has begun. Virginia legislators have proposed bills requiring data centers to pay a larger share of infrastructure costs. Consumer advocates argue that data center operators—billion-dollar companies serving other billion-dollar companies—should not receive subsidies from residential ratepayers earning median incomes. The counterargument: data centers create jobs and tax revenue that benefit everyone.

The debate reflects a fundamental question about who pays for infrastructure that primarily benefits private companies. Data centers generate enormous value for their owners and customers—value that flows to shareholders and corporate users. But the costs of enabling that value are spread across all ratepayers. This represents a transfer from households to corporations, mediated by the regulated utility system.

The Saline Township project navigated these concerns by structuring power contracts to isolate costs. According to DTE and the Michigan Public Service Commission, if the data center fails to pay, DTE cannot pass costs to other customers. The contract supposedly creates a $300 million "net benefit" flowing to other ratepayers. But with key provisions redacted as proprietary, Michigan's Attorney General Dana Nessel noted that verifying these claims is "impossible."[93]

The opacity of data center power contracts makes public oversight difficult. Utilities and their large customers negotiate terms in private, present agreements to regulators for approval, and redact the details that would allow meaningful analysis. The pattern raises a question: are consumer protection provisions genuine, or merely cosmetic?

Industry advocates offer a counterargument: data centers should *reduce* electricity costs by spreading the grid's fixed costs across more kilowatt-hours sold. A 2024 Lawrence Berkeley National Laboratory study found that load growth correlated with approximately 0.6 cents per kilowatt-hour reduction in average retail prices.[94] But the researchers embedded critical caveats: "This relationship need not always exist: a higher growth future can increase retail prices if new supply and delivery infrastructure is constrained and costly."

The current buildout involves gigawatt-scale single facilities, severe infrastructure constraints, and massive capital requirements—exactly the conditions where cost-sharing logic breaks down. When infrastructure must be *built* rather than merely *used*, new capacity costs exceed the revenue benefits. PJM's capacity market saw prices rise from $28.92 per megawatt-day in 2024 to $269.92 in 2025—a ninefold increase driven substantially by data center load growth.[95] The Market Monitor attributed 63 percent of the price increase to data center demand, translating to $9.3 billion in additional costs.[96] Meanwhile, data centers typically receive discounted industrial rates—often 30 to 50 percent below residential—contributing less per kilowatt-hour to fixed cost recovery.

The policy response confirms that legislators do not believe cost-sharing benefits materialize automatically. Texas Senate Bill 6, Oregon's POWER Act, and Virginia's new large-load rate class exist because policymakers recognized that without intervention, residential ratepayers subsidize data center infrastructure.

The honest assessment: cost-sharing benefits *might* materialize in the long term, once infrastructure is built and data centers contribute their contracted amounts over decades. But in the short term, residential ratepayers face documented cost increases. The question is not whether data centers eventually pay for themselves—it is who bears the cost and risk during the transition.

5.10 FERC ORDER 2023

Recognizing the queue crisis, the Federal Energy Regulatory Commission issued Order 2023 in July 2023—the most significant interconnection reform in two decades.[97] The order took effect that November, with compliance required by April 2024.

The reforms address several problems. First, the order shifts from sequential first-come-first-served studies to clustered studies that evaluate groups of projects together, reducing redundant analysis and speeding the process. Instead of studying each project in isolation, ISOs can identify common infrastructure needs across multiple projects.

Second, the order increases financial commitments required from project developers. Higher deposits and milestone payments should discourage speculative applications that clog the queue without serious intent to build. A developer putting down $100,000 thinks carefully about whether the project is real. A developer putting down $10,000 might apply speculatively.

Third, the order establishes binding timelines for study completion. ISOs must complete cluster studies within specific windows or face penalties, preventing indefinite delays. Projects should know within two years whether they can proceed, rather than waiting indefinitely for study results.

Fourth, the order creates affected system study requirements, ensuring that projects near ISO boundaries receive coordinated review. A project affecting both PJM and MISO should not require separate, uncoordinated studies in each region.

The reforms represent genuine progress, but their impact remains limited. The order does not address the underlying shortage of transmission infrastructure, does not accelerate permitting of new transmission lines, does not resolve staffing constraints at ISOs struggling to process thousands of applications. It optimizes a process without expanding the underlying capacity.

For data centers specifically, the reforms may provide marginal improvement in study timelines without fundamentally changing the calculus. A project facing an eight-year wait might now face six. For companies that need power in eighteen months, neither timeline works.

In October 2025, the Department of Energy urged FERC to consider additional reforms specifically addressing large-load interconnection, including faster procedures for data centers.[98] The letter acknowledged that data center growth represents a novel challenge existing processes were not designed to handle.

Whether FERC will act, and how quickly, remains uncertain. The commission moves deliberately—new rulemakings take years. The data center industry cannot wait for regulatory solutions that might arrive in 2028 or 2030.

5.11 GEOGRAPHIC CONSEQUENCES

The constraints described in this chapter—queue backlogs, transmission limitations, the farmland paradox—are not merely technical problems. They shape the geography of America's digital economy.

Data centers cluster where grid capacity exists. Northern Virginia became Data Center Alley because Dominion Energy had capacity to serve the original facilities, the fiber networks were already there, and early success attracted later investment. But the region is now saturated. New projects push outward to rural counties—Louisa, Culpeper, Spotsylvania—where transmission infrastructure can still accommodate growth.

The expansion creates new tensions. Rural Virginia counties have discovered that proximity to Data Center Alley means pressure for development their residents did not anticipate. A county that was primarily agricultural suddenly faces proposals for industrial-scale facilities. The tax revenue is attractive; the transformation, unsettling. Local governments must decide whether to embrace the change or resist it.

Texas has grown rapidly because ERCOT's fast-track processes accommodate large loads that would wait years elsewhere. The risk of grid instability has not deterred developers facing worse alternatives. Dallas-Fort Worth has become a major data center market; Austin and San Antonio are growing. The vast spaces of West Texas, with abundant wind resources and minimal population density, have attracted projects that would face opposition in more developed areas.

The Midwest is emerging as a new frontier. Michigan, Ohio, Indiana, and Wisconsin all report surging data center interest. MISO's interconnection process, while slow, is less congested than PJM's. State incentive programs have grown aggressive, and land remains abundant and cheap. A company that cannot build in Virginia might build in Indiana instead.

The geographic shift has implications beyond the data center industry itself. Where data centers go, tech jobs follow—not the thousands promised in press releases, but the hundreds of skilled positions that actually materialize. Tax revenue flows to local jurisdictions, creating winners and losers among neighboring communities. Power consumption reshapes regional grids, affecting reliability and rates for all customers.

For rural communities like Saline Township, these changes arrive suddenly and at enormous scale. A township of 2,300 people receives a proposal for a 1.4-gigawatt facility. The numbers are incomprehensible at the local level—the investment exceeds the township's total tax base many times over, the power consumption exceeds anything the community has ever experienced. The decisions are made by actors operating at state, national, and global scales. The community can accept or resist, but it cannot fundamentally alter the forces driving the transformation.

5.12 Looking Forward

The grid constraints described in this chapter are not temporary problems awaiting easy solutions. They reflect decades of underinvestment in transmission infrastructure, regulatory fragmentation that slows new construction, and a planning process designed for incremental growth—not step-function increases in demand.

Addressing these constraints will require coordinated action across multiple levels of government. Federal reforms can accelerate interconnection studies but cannot force states to approve new transmission lines. State regulators can simplify permitting but cannot solve the financial challenges of building infrastructure that crosses state lines. Utilities can invest in upgrades but require regulatory approval and cost recovery mechanisms that take years to negotiate.

The data center industry, meanwhile, will continue adapting. Behind-the-meter generation will expand as queue backlogs persist. Nuclear power partnerships, discussed in the next chapter, represent a bet on long-term solutions to the baseload power problem. Geographic diversification will spread facilities across regions with available capacity.

The fundamental tension remains unresolved: data centers need power measured in gigawatts, on timelines measured in months, from a system designed to provide power measured in kilowatts on timelines measured in years. Something must give. Either the grid must change to accommodate new demands, or the demands must change to fit existing infrastructure.

The Saline Township project represents one answer: a location with existing transmission capacity, a utility willing to prioritize service, and a project large enough to finance its own infrastructure. But transmission corridors crossing farmland are finite. The model cannot accommodate a hundred more gigawatt-scale projects.

The grid we have was built for a world that no longer exists. The grid we need does not yet exist. The gap between them defines what data centers can achieve—and understanding that gap is the first step toward either transcending those limits or accepting them.

We turn now to where the power itself comes from—the generation sources that feed the grid and the facilities that bypass it. The grid determines where data centers can connect. Generation determines the carbon and environmental footprint of what they do.

CHAPTER SIX

The Generation

THE CONTROL ROOM AT THREE MILE ISLAND Unit 1 had been dark for nearly a decade when *Elena* walked back in. September 2019 was when Exelon had shut down the reactor, unable to compete with cheap natural gas. Now, on a morning in late 2028, she stood before the same control panels where she had spent a decade of her career, watching as technicians performed the startup sequence for Microsoft.

The irony was not lost on her. Three Mile Island—the name synonymous with nuclear fear since Unit 2's partial meltdown in 1979—would power artificial intelligence. The same site where America's worst commercial nuclear accident had nearly destroyed the industry would now generate the carbon-free electricity that technology companies had promised to their shareholders.

"Criticality in fifteen minutes," the shift supervisor announced. *Elena* watched the instrumentation. Neutron flux rising. Control rods withdrawing. The reactor coming back to life after years of dormancy.

She thought about the journey that had brought her here: Constellation Energy's announcement in 2024, the twenty-year power purchase agreement with Microsoft, the years of regulatory review and equipment upgrades.[99] She thought about the competing announcement she had read the week before—a massive gas plant in western Pennsylvania that would generate 4.5 gigawatts for AI data centers.[100] Nuclear plants take decades to restart. Gas plants take years. The AI boom operates on time-lines measured in months.

The reactor went critical at 8:47 AM. The first electrons flowed toward Microsoft's data centers. *Elena* allowed herself a moment of satisfaction. But she knew this was the exception. Across America, most of the power feeding AI came from burning fossil fuels—and would for years to come. This chapter traces where that power actually comes from, the gap between sustainability promises and operational reality.

6.1 THE AMERICAN GENERATION MIX

To understand data center power, we need to understand American electricity. The electrons that light homes and run factories are the same electrons training large language models. Data centers draw from the grid, and the grid reflects a particular mix of generating sources.

As of 2025, natural gas provides about 40 percent of American electricity, making it the dominant fuel source. Low-carbon sources collectively provide about 43 percent. Nuclear contributes roughly 17 percent. Wind adds about 10 percent. Solar provides 8 percent. Hydroelectric rounds out with 6 percent. Coal has declined to about 16 percent.[101] Gas dominant, low-carbon rising, coal retreating. This mix forms the baseline against which every data center's power claims must be measured.

This mix has transformed over two decades. In 2005, coal provided half of American electricity. Then the shale gas revolution collapsed coal's economics, and environmental regulations accelerated the shift. Today, wind and solar generate more electricity than coal.[102]

Regional variation shapes what powers specific facilities. In Washington state, a data center might draw primarily from hydroelectric dams. In Texas, another might run on wind, solar, and natural gas in a mix that shifts hour by hour. In West Virginia, coal and gas dominate. Location matters not just for interconnection speed, as we saw in the previous chapter, but for the carbon intensity of every computation performed.

The grid's marginal generation—the power plant that ramps up when a new load comes online—is almost always natural gas. When a data center switches on and demands 500 megawatts of continuous power, somewhere a gas turbine spins faster. The efficiency paradox we examine later shows why this matters more than corporate sustainability reports suggest.

6.2   THE NATURAL GAS REALITY

Natural gas has become the default fuel for new data center power, despite corporate sustainability commitments that suggest otherwise. The numbers are stark: of the eleven multi-gigawatt data center projects announced through late 2024, ten rely on natural gas as their primary power source.[103]

The reasons are straightforward. Gas turbines are proven technology that deploys quickly—two to three years from decision to operation. They provide reliable baseload power, the constant electricity a grid must deliver around the clock regardless of weather or time of day. Developers can build them behind the meter, on the data center's own property, bypassing interconnection queues that add years to grid-connected projects. And the economics work: $40 to $70 per megawatt-hour, competitive with anything else.

The technology itself is mature and well-understood. A combined-cycle gas turbine plant uses two stages to extract energy from natural gas. First, combustion gases spin a turbine directly. Then, the hot exhaust raises steam to spin a second turbine. Combined-cycle plants achieve thermal efficiencies of 60 percent or more, meaning they convert 60 percent of the fuel's energy content into electricity. Simple-cycle plants, which skip the steam stage, achieve lower efficiencies but can ramp up and down more quickly. Both designs are manufactured by established suppliers like GE Vernova, Siemens, and Mitsubishi.

Western Pennsylvania has become the epicenter of gas-powered data center development. The region sits atop the Marcellus Shale, one of the largest natural gas deposits in the world. Three massive projects have emerged there in rapid succession.

Homer City Energy Center plans 4.5 gigawatts on the site of a former coal plant, powered by seven GE Vernova gas turbines.[104] The coal plant that once stood here was among the largest in the eastern United States; it closed in 2023 after decades of operation. Now the site will host natural gas generation at even larger scale. TECfusions Keystone Connect plans 3.0 gigawatts at a former Alcoa research campus, combining natural gas with hybrid renewable options. The Shippingport project plans 2.7 gigawatts where a nuclear reactor once operated, now fueled by Marcellus gas.

Together, these three projects represent more than 10 gigawatts of gas-powered data center capacity in a single state. Gas flows through pipelines running beneath the very servers that will consume it. Pennsylvania's energy history—coal to nuclear to gas—plays out in these sites.

Similar patterns emerge elsewhere. In Utah, the Delta Gigasite (4.0 gigawatts) and the Joule Capital project (4.0 gigawatts) rely on natural gas for baseload power; proximity to western gas fields provides fuel supply security. In Arizona, the Vermaland La Osa project (3.0 gigawatts) plans to start with gas before transitioning to solar. The desert offers abundant sunshine but cannot provide power at night without storage that remains expensive. In West Virginia, the Adams Fork Energy project spreads 4.8 gigawatts across two sites, combining gas with carbon capture and ammonia production.

The environmental implications are substantial. A single 500-megawatt data center running on natural gas emits roughly 2 million tons of carbon dioxide per year—equivalent to 400,000 cars on the road. Scale that to the industry's projected growth, and the numbers become alarming. Goldman Sachs estimates that data center expansion will add

200 million tons of annual carbon emissions by 2030.[105] By the mid-2030s, according to the Washington Post, data center emissions could equal those of New York City, Chicago, and Houston combined.[106]

Natural gas also releases methane during extraction, processing, and transport. Methane traps roughly 80 times more heat than carbon dioxide over a twenty-year period. The leakage rate from natural gas systems—estimated at 2 to 3 percent of production—significantly increases the true climate impact of gas-fired generation. When methane leakage is included, natural gas loses much of its advantage over coal.

But carbon is not the only impact. Water consumption tells its own story, and nowhere more sharply than in Arizona.

Roberto farms 800 acres of alfalfa outside Phoenix. His grandfather came from Sonora in 1952, worked as a bracero for six years, saved enough to buy forty acres of desert that nobody wanted, and spent thirty years proving what water could do to sand. His father expanded to 400 acres. *Roberto* doubled it again. Now he is sixty-one, his knees ache from decades on a tractor, and he can feel the ground drying beneath him.

Three generations of water rights. Three generations of knowing exactly how much the Salt River Project would deliver each year, planning crops around it, timing harvests to the week. The allocation was as reliable as the sunrise.

In 2024, that allocation dropped 20 percent—the largest cut in his lifetime.

Roberto stands in a field that used to grow four cuttings of alfalfa per year and now grows three. The stubble crunches under his boots. Alfalfa needs four acre-feet of water per acre per year in this climate. He can do the math faster than any accountant: twenty percent less water means twenty percent less crop, means twenty percent less income, means decisions he never thought he would have to make.

"They told us agricultural users had to conserve," he says. His pickup truck idles nearby, its bed stained green from decades of hauling feed.

"Meanwhile, they're approving water rights for data centers like it grows on trees." The raw numbers favor the data centers—a gigawatt facility uses far less water annually than *Roberto*'s 800 acres. But hydrological reality is more complicated. Agricultural water percolates through soil, recharges aquifers, returns through plant transpiration to fall as rain downwind. Evaporative cooling sends water into the atmosphere as pure vapor, often lost to the local watershed entirely. The acre-feet comparison obscures these differences. *Roberto*'s sense of unfairness may be more defensible than it first appears.

He points toward the horizon, where a new development is visible as a smudge of construction dust. "You know what they cool those things with? Evaporation. Same water my grandfather used to turn this desert into fields. They spray it on machines, it turns to steam, it's gone. At least when I use water, it grows something. It feeds cattle. It feeds people."

Roberto learned about water in ag school at Arizona State, back when his father thought a college education would make him a better farmer. He still remembers the lesson that stuck: every drop of water in a crop field participates in a cycle. The roots pull it up; the leaves release it slowly through transpiration; some of it falls back as dew; some returns to the aquifer. Not all of it, not most of it—but some. Irrigation is inefficient, sure. Maybe sixty percent of what he pumps actually reaches the roots. But the rest does not vanish. It seeps back into the ground. Evaporative cooling is different. Water hits the hot metal, flashes to vapor, rises into the sky, and drifts away on the wind. Gone to the atmosphere. Gone to Nevada, maybe, or California. The desert does not get it back.

The math is straightforward, if uncomfortable. *Roberto*'s 800 acres consume roughly 3,000 acre-feet of water annually—enough for perhaps 9,000 households. A single gigawatt-scale data center using evaporative cooling can consume 1,000 acre-feet per year. Arizona has approved water allocations for data center projects that collectively exceed 50,000 acre-feet annually. Every acre-foot flowing to data centers is an acre-foot not available for farms.

His daughter *Maria* works in Phoenix now, in an office building with air conditioning that runs all summer. She tells him the future is technology, not agriculture. His grandson plays video games on a tablet. *Roberto* wonders if the boy has ever seen an alfalfa field at dawn, when the irrigation water catches the first light and the whole world seems to shimmer.

"I don't begrudge anybody making a living," *Roberto* says. "But when they tell me to fallow fields my grandfather planted, while a company from California gets water rights for cooling servers? That's not conservation. That's a transfer."

He looks out at his remaining fields. The ones he might have to sell. The ones his grandson will never farm.

"My grandfather came here with nothing. Built something from nothing. Now they're telling me his grandkids will inherit nothing again." He kicks at the dry soil. "Different kind of nothing, I guess. But nothing all the same."

The Tucson city council voted unanimously in 2025 to reject an Amazon data center proposal over water concerns.[63] Community opposition killed the $250 million project. But for every Tucson, a dozen proposals proceed. The water table keeps dropping. The trade-offs keep happening, mostly invisibly, while corporate sustainability reports celebrate solar panels and ignore the wells running dry.

These environmental costs—carbon emissions, methane leakage, water consumption—create the sustainability paradox. The same companies building these gas-fired facilities are the ones making the most aggressive renewable energy commitments. The press releases promise carbon neutrality. The permit applications request fossil fuel infrastructure. Both statements are technically true, but they operate on incompatible timescales.

6.3 CARBON ACCOUNTING

How can technology companies build natural gas plants while claiming renewable energy progress? The answer lies in corporate carbon accounting, which is more permissive than most people realize.

Corporate emissions are divided into Scope 1 (direct emissions from company-owned sources), Scope 2 (indirect emissions from purchased electricity), and Scope 3 (everything else in the value chain, from server manufacturing to employee commuting). Most corporate renewable energy claims focus on Scope 2, addressed primarily through renewable energy certificates, or RECs.

When a solar farm generates a megawatt-hour of electricity, it creates two products: the physical electron flowing into the grid, and a REC certifying that renewable generation occurred. These products can be sold separately. A data center in Virginia might draw electrons from a grid that is 60 percent natural gas, but if the company buys RECs from a Kansas wind farm equal to its annual consumption, it can claim "100 percent renewable" in its sustainability report. The electrons are no cleaner. The wind farm was going to generate that power anyway. But the accounting shows green.

This practice, called "annual matching," allows purchases from any region or even different years. Amazon, which claims 100 percent renewable energy as of 2025, relies heavily on it.

The problem is that annual matching does nothing to ensure that actual renewable power meets actual data center demand at the moments when that demand occurs. A data center drawing 500 megawatts at midnight cannot use solar power generated at noon. The REC accounting pretends the two are equivalent. The atmosphere knows better.

Google has pioneered a different approach called "24/7 carbon-free energy" or "hourly matching"—aligning electricity consumption with carbon-free generation on an hourly basis, in every region where it operates. This is far more demanding. It requires either storage to shift renewable generation to nighttime hours, or baseload carbon-free power operating around the clock.

This is why nuclear energy has become so attractive to hyperscalers. Nuclear plants produce steady power day and night, regardless of weather, running at 90 percent or higher capacity factors—meaning they generate power at or near their maximum output for nine out of every ten hours. They are the only proven technology that can deliver 24/7 carbon-free baseload at scale. Microsoft has adopted similar 24/7 goals, which helps explain its willingness to pay premium prices to restart Three Mile Island.

The gap between annual matching and hourly matching reveals the gap between marketing and physics. A company can claim "100 percent renewable" while its data centers run primarily on natural gas.

A full accounting of data center carbon impacts—embodied emissions in construction materials, Scope 3 emissions from the AI supply chain, the atmospheric science of methane leakage—would require another book. This chapter focuses on the gap between corporate claims and operational reality, not on quantifying total climate impact. The same holds for water: competition between data centers and agriculture for scarce groundwater in the American West is tangled in century-old rights, interstate compacts, and tribal claims that we can only gesture toward here.

6.4 THE EFFICIENCY PARADOX

A common industry defense holds that blocking data center construction actually *hurts* the environment. The argument goes like this: hyperscale facilities are far more efficient than enterprise data centers, achieving power usage effectiveness ratios of 1.1 to 1.2 compared to 1.5 to 1.8 for typical corporate facilities.[107] If compute demand exists, better to concentrate it in efficient hyperscale facilities than scatter it across inefficient enterprise servers. Therefore, opposing new hyperscale construction pushes computation to dirtier alternatives.

The efficiency numbers are accurate. Google's data centers use approximately 84 percent less overhead energy per unit of IT equipment than the industry average; AWS and Microsoft achieve similar results. These companies have spent billions optimizing their facilities because electricity is their largest operating cost. The incentives align.

But the argument contains a flaw: it assumes that blocking a hyperscale facility merely relocates computation. The more likely outcome is some combination of relocation, substitution, and demand destruction—the last of which would reduce emissions, not increase them.

When AWS withdrew its Project Blue proposal from Tucson in 2025, the computation that would have occurred there did not simply materialize elsewhere at lower efficiency. Some demand may have relocated to other AWS facilities. Some may have been served by expanding existing data centers incrementally. And some—speculative projects, marginal applications, experiments that would not have occurred at higher cost—may never have materialized at all.

The assumption that compute demand is perfectly inelastic—that every gigawatt of blocked capacity would be built elsewhere—reflects wishful thinking rather than market analysis. In China, hundreds of AI data centers now sit unused because "the push to build AI data centers was largely driven from the top down, often with little regard for actual demand," according to industry analysts.[108] This speculative overbuilding suggests that not all proposed capacity would generate equivalent demand if blocked.

Even setting aside the demand question, the efficiency argument confronts the Jevons paradox introduced in Chapter 2: efficiency gains can enable demand growth that overwhelms per-unit savings.

Applied to AI data centers, the paradox suggests that building efficient infrastructure enables demand growth that overwhelms efficiency gains. A facility that is 30 percent more efficient but serves 10 times more computation produces 7 times the emissions. Google's own data illustrates this: despite industry-leading efficiency, the company's data center energy consumption has grown dramatically. Its 2024 environmental report acknowledged that AI had increased emissions by 48 percent since 2019, even as efficiency improved.[109]

The most fundamental problem with the efficiency defense is that it ignores *what power source* new data centers actually use. Efficiency per computation means nothing if total computation grows and the incremental power comes from fossil fuels.

The evidence is troubling. Natural gas generation for data centers is projected to more than double, from 120 terawatt-hours in 2024 to 293 terawatt-hours in 2035.[17] Meta has entered agreements with Entergy to construct three new natural gas plants in Louisiana. OpenAI's Stargate project includes a 900-megawatt gas plant. Virginia approved Dominion's first new gas plant since the Clean Economy Act specifically to serve data center demand.[110] At least seventeen fossil fuel generators originally scheduled for closure have delayed retirement because of data center demand.[111]

The marginal emissions problem explains why average grid mix statistics mislead. A grid might be 40 percent renewable on average, but new data center demand at midnight is met by whichever generators can ramp up—typically natural gas. A 2022 PNAS study found that while average emissions from US electricity declined 40 percent between 2006 and 2019, marginal emissions—the emissions from adding new demand—declined only 8 percent.[112] Building efficient data centers powered by new gas plants increases emissions regardless of facility PUE.

None of this means that blocking every data center proposal benefits the environment. The honest assessment is that environmental outcomes depend on specific circumstances. What power source would the blocked facility have used? Does demand relocate or disappear? Is the alternative more or less efficient? Do AI applications enabled by the infrastructure generate emissions savings elsewhere in the economy? Each question matters. The answers vary by project.

The International Energy Agency estimates that widespread adoption of AI applications could yield substantial emissions reductions—perhaps exceeding data center emissions themselves—through improved industrial processes, grid optimization, and building efficiency.[17] If AI enables a factory to cut energy consumption by 20 percent, the emissions savings might exceed the emissions from powering the AI. But this calculus depends on whether AI is actually deployed for efficiency applications rather than inference for consumer entertainment, cryptocurrency, or other uses with limited emissions benefits.

The debate cannot be resolved with simple heuristics. "Efficient, therefore good" ignores the Jevons paradox, marginal emissions, and the specific power sources serving new demand. "Energy-intensive, therefore bad" ignores the potential for AI to enable emissions reductions in other sectors. The answer depends on the particulars, and the particulars vary by project, by region, and by what applications the infrastructure ultimately serves.

What we can say with confidence is that the efficiency argument, while factually accurate about PUE comparisons, does not establish that building more data centers helps the environment. Total emissions continue rising. New fossil fuel plants are being built specifically to serve data center demand. Corporate renewable claims rely heavily on accounting mechanisms that do not reduce actual emissions. The gap between aspiration and reality defines this moment in the industry's history.

6.5 THE NUCLEAR RENAISSANCE

In late 2024, something remarkable happened. Within weeks of each other, three major technology companies announced nuclear power partnerships: Microsoft with Constellation Energy to restart Three Mile Island, Google with Kairos Power to develop small modular reactors,[113] and Amazon with X-energy to build a fleet of advanced reactors.[114] After decades in which nuclear seemed politically untouchable, the technology industry embraced atoms with something like religious fervor.

The timing was not coincidental. Each company had committed to 24/7 carbon-free energy goals they could not meet with renewables alone. Each faced scrutiny over the carbon intensity of their AI operations. And each had the balance sheet strength to make decade-long bets on unproven technology. Nuclear offered something nothing else could: continuous carbon-free generation at scale.

These partnerships represent more than 24 gigawatts of committed capacity—roughly equivalent to twenty-four large reactors, enough power for 20 million American homes. If all these projects proceed as planned, they will transform both the nuclear industry and the data center sector. But the timeline stretches to 2044 and beyond, and the path from announcement to operation is littered with obstacles.

———

The most immediate project is restarting Three Mile Island Unit 1, the undamaged reactor adjacent to the infamous Unit 2. The 1979 partial meltdown at Unit 2 became the defining event of American nuclear history, halting new reactor construction for a generation. But Unit 1, separated from Unit 2 by only a few hundred meters of Pennsylvania farmland, continued operating safely for decades.

Unit 1 shut down in 2019 when low natural gas prices made it uneconomical. Exelon, then the owner, had lobbied Pennsylvania for subsidies to keep the plant running. When subsidies failed to materialize, the company mothballed the reactor rather than decommissioning it. Mothballing preserved the option to restart; decommissioning would have been irreversible.

Constellation Energy, which acquired the plant when it spun off from Exelon in 2022, has now exercised that option. Microsoft's twenty-year power purchase agreement provides the revenue certainty to justify restart costs. The deal guarantees Microsoft access to the plant's entire output: 835 megawatts of continuous carbon-free generation, worth roughly $300 million per year at retail electricity rates.

Constellation initially estimated $1.6 billion for the restart, with operations beginning in 2028. The work requires relicensing from the Nuclear Regulatory Commission, rehiring and retraining operating staff, and refurbishing equipment that has sat idle for years. Steam generators, turbines, cooling systems, control room equipment—all need inspection and repair.

In November 2025, the Department of Energy provided a $1 billion loan guarantee, recognizing the project's significance for both energy policy and AI competitiveness.[115] Constellation accelerated the timeline to 2027. The company has achieved 65 percent staffing and begun equipment refurbishment.

Constellation renamed the facility the "Crane Clean Energy Centre"—after Chris Crane, the former Exelon CEO who preserved the restart option—hoping to distance it from 1979. Whether the rebranding succeeds, the restart itself represents the largest addition of carbon-free generation to the American grid since 2016: 835 megawatts. It demonstrates that existing nuclear plants, even those shuttered for years, can return to life when economic conditions change. And it establishes Microsoft as the leading hyperscaler in nuclear procurement.

Meta followed with its own Constellation deal in June 2025: a twenty-year agreement for power from the Clinton nuclear plant in Illinois.[116] The 1,121-megawatt reactor will begin delivering power to Meta facilities in June 2027. Between TMI and Clinton, Constellation has contracted nearly 2 gigawatts of nuclear capacity to technology customers, becoming the nuclear industry's primary link to the AI boom.

———

Beyond restarts, the nuclear renaissance centers on small modular reactors (SMRs)—factory-built units that promise lower costs and faster deployment than traditional large reactors. The technology has operated abroad: China's HTR-PM began generating power in 2023, and Russia has deployed floating reactors in the Arctic. But American SMR developers remain in earlier stages—working through design certification, construction permits, and demonstration projects before commercial deployment can begin.

Traditional nuclear reactors are massive, custom-built projects. A single reactor might produce 1,000 megawatts or more, requiring a decade of construction and $20 billion or more in capital. The custom nature of each project means lessons from one do not transfer to the next. Every reactor is, in some sense, first of its kind.

SMRs aim to change this. They produce 50 to 300 megawatts per unit, small enough to manufacture in factories and ship to sites. Factory production enables standardization: each reactor identical to the last, allowing workers to climb learning curves and costs to decline over time. Modular deployment allows incremental capacity addition—a customer can start with one reactor and add more as demand grows.

Multiple competing designs exist within the SMR category, using different fuels and coolants—molten salt, high-temperature gas, liquid sodium. The politics and science of restarting American nuclear development are genuinely complex. What follows reflects that complexity: different companies pursuing different technologies at different stages of regulatory approval, with timelines stretching years into the future.

Google's partnership with Kairos Power focuses on molten salt reactors, a design that uses liquid fluoride salt instead of water as coolant. Molten salt designs offer several advantages: they operate at atmospheric pressure, eliminating the high-pressure piping that makes traditional reactors complex; they run at higher temperatures, improving efficiency;

147

and they cannot melt down in the conventional sense, because the fuel is already molten.

Kairos has reached the demonstration stage. Hermes, a small test reactor that will not generate commercial power, received NRC construction approval in December 2023.[117] The larger Hermes 2 was approved in November 2024. Both are being built in partnership with the Tennessee Valley Authority. In August 2025, Google and Kairos announced a deal with TVA to connect Hermes 2 to the utility grid—the first commercial power purchase agreement for a Generation IV advanced reactor, though at only 50 megawatts.[118] Commercial-scale deployment remains years away: Google has contracted for 500 megawatts of capacity, with delivery targeted between 2030 and 2035.

Amazon's partnership with X-energy targets even larger scale. X-energy's Xe-100 is a high-temperature gas-cooled reactor that uses TRISO fuel—tiny uranium particles coated in ceramic layers that can withstand extreme temperatures. TRISO fuel is sometimes called "nuclear fuel you can hold in your hand": the ceramic coatings trap radioactive materials inside, preventing release even if the reactor fails catastrophically. The design is inherently safe in ways that earlier reactor designs were not.

Through its X-energy investment, Amazon targets 5 gigawatts of SMR capacity by 2039—roughly equivalent to five traditional large reactors. The first deployment will be the Cascade facility in Washington state: twelve Xe-100 reactors totaling about 1 gigawatt, built in partnership with Energy Northwest.[119] Amazon has invested approximately $500 million in X-energy and taken a board seat to ensure close coordination.

Switch, the data center operator based in Las Vegas, announced the largest nuclear commitment: a memorandum of agreement with Oklo for 12 gigawatts by 2044.[120] Oklo builds Aurora reactors—small sodium-cooled fast reactors that can run on spent fuel from conventional reactors. The Aurora design produces 15 to 50 megawatts per unit, meaning the 12-gigawatt commitment would require hundreds of reactors. The timeline extends twenty years into the future, and the agreement is non-binding—expressing intent rather than commitment.

Oracle has announced plans for 1,000 megawatts of SMR capacity to power a data center, with permits reportedly secured. Larry Ellison, Oracle's co-founder and chairman, has spoken enthusiastically about nuclear-powered AI facilities. Details remain sparse, but the project reflects the industry-wide pivot toward nuclear.

The nuclear announcements have generated enormous media attention and genuine excitement. But we should be clear-eyed about timelines. The history of nuclear construction is one of delays and cost overruns.

The most recent American nuclear project—Vogtle Units 3 and 4 in Georgia—illustrates the challenge. These reactors were supposed to cost $14 billion and begin operating in 2016 and 2017. They actually cost more than $35 billion and began operating in 2023 and 2024.[121] The delays bankrupted Westinghouse, the lead contractor. Traditional nuclear construction remains extraordinarily difficult in the American regulatory and labor environment.

Small modular reactors promise to break this pattern through factory manufacturing and standardized designs. But the only SMR design the NRC has approved for construction—NuScale's Voygr[122]—saw its flagship deployment at Idaho National Laboratory cancelled in 2023 when costs exceeded projections.[123] SMR economics remain unproven.

A realistic timeline for SMR deployment looks like this: The first utility-connected demonstration reactor (Hermes 2) might operate by 2030. Commercial-scale deployments might follow by 2032 to 2035. Scale deployment, with multiple reactors operating and costs declining along learning curves, might arrive by 2035 to 2040. The 12 gigawatts Switch promised with Oklo? That is a 2040s project at best.

Meanwhile, AI demand operates on six-to-eighteen-month cycles. A company that needs power today cannot wait for reactors that will not operate for a decade. This is why natural gas dominates near-term construction even as nuclear dominates long-term planning. The gap between announcement and operation is filled with fossil fuel.

Given the cost uncertainty, regulatory complexity, and decade-long timelines, why pursue nuclear so aggressively?

Because it is the only proven technology for 24/7 carbon-free baseload power at scale. Hydroelectric cannot expand significantly; the best dam sites are already developed. Solar and wind require storage to provide continuous power. Only nuclear can deliver hundreds of megawatts of carbon-free power continuously, anywhere.

Geothermal deserves a brief aside. Traditional geothermal works only where hot rock lies close to the surface—Iceland, the Geysers in California, scattered sites in Nevada and Utah. But enhanced geothermal systems (EGS) drill deeper and inject water to fracture rock, creating reservoirs where none existed naturally. Google's partnership with Fervo Energy brought the first commercial EGS plant online in Nevada in 2024, delivering 3.5 megawatts to the grid.[124] Companies like Zanskar are using AI to identify viable sites beyond volcanic regions, potentially expanding geothermal's geographic reach.[125] The technology offers 24/7 baseload power with minimal land footprint and no intermittency. But scale remains the challenge: Fervo's entire Nevada output would power a single row of AI racks. Geothermal may eventually contribute meaningful capacity, but not at the gigawatt scale data centers require today.

Nuclear plants also have small physical footprints: 1 to 2 square miles for 1,000 megawatts, versus 10 to 15 square miles for equivalent solar. They last sixty to eighty years, compared to twenty or thirty for renewables. Their stable operating costs help data centers forecast power expenses over decades. And if carbon prices rise through cap-and-trade or carbon taxes, nuclear's zero-emission profile becomes economically advantageous. Technology companies are betting that carbon will eventually cost something.

6.6 RENEWABLES

While natural gas dominates near-term construction and nuclear dominates long-term planning, renewables play an important and growing role. In the first ten months of 2025, renewables provided about 25.7 percent of American electricity.[126] Costs have plummeted: utility-scale solar runs roughly $30 to $50 per megawatt-hour, wind $20 to $40—cheaper than any fossil fuel alternative in most markets.[127]

These favorable economics have attracted data center investment. Of the 604 documented projects in my database, 149 (25 percent) have explicit renewable energy components, representing 49 gigawatts (37 percent) of planned capacity.[8] Oregon's data centers run on nearly 100 percent carbon-free power, thanks to Columbia River hydroelectric dams. Iowa and Wyoming approach 70 percent renewable power for their data center sectors.

Power purchase agreements (PPAs) have become the primary vehicle for data center renewable procurement. Under a PPA, a data center operator contracts with a renewable generator to purchase its output at a fixed price over ten to twenty years. Amazon has become the world's largest corporate purchaser of renewable energy. These agreements have directly funded gigawatts of new wind and solar capacity that might not otherwise have been built.

The challenge is intermittency. Solar panels produce nothing at night; wind turbines depend on weather. A data center that needs 500 megawatts continuously cannot rely solely on sources that fluctuate between 0 and 150 percent of rated capacity.

Battery storage partially addresses intermittency, but at significant cost. Storing four hours of a 500-megawatt load requires 2,000 megawatt-hours of battery capacity, costing roughly $500 million at current prices. Storage adds $20 to $40 per megawatt-hour to renewable electricity costs. A data center cannot reduce consumption when the wind stops blowing. AI training jobs cannot pause until the sun comes out. Meeting constant load requires either fossil backup or storage at a scale that remains prohibitively expensive.

This reality shapes hyperscaler renewable strategies. Companies pursue power purchase agreements with renewable generators while remaining connected to grids that provide backup. Some projects attempt hybrid approaches: Vermaland La Osa in Arizona plans to start with natural gas while building out solar capacity over time; Data City Texas promises 5 gigawatts transitioning from gas to hydrogen. These are transition strategies—acknowledgments that pure renewables cannot yet meet data center needs.

The most honest assessment is that renewables will power a growing share of data center electricity, but not the majority share, and not on the timelines that corporate sustainability reports imply. Natural gas will fill the gap in the near term; nuclear may fill it in the long term, if the technology delivers.

6.7 Saline's Power

The Saline Township data center will consume 1.4 gigawatts—between fifteen and twenty-five percent of DTE Energy's current peak capacity, depending on how peak load is measured. Where will that power come from? The answer reveals the gap between aspiration and reality.

DTE Energy's current generation mix includes significant fossil fuel capacity. The utility operates multiple natural gas plants at Belle River, Monroe, and River Rouge. It operates Fermi 2, a nuclear reactor providing about 1,100 megawatts of baseload capacity. It is building renewable capacity to meet Michigan's clean energy mandates, including several hundred megawatts of wind and solar under development. But the current mix remains heavily fossil-dependent—roughly 60 percent of DTE's generation comes from fossil fuels.

DTE claims it can serve the facility from existing resources augmented by battery storage, without building new generation—a remarkable assertion given normal utility capacity planning, which typically maintains 15 to 20 percent reserves above peak demand.

More likely, serving Saline will require a combination of strategies: purchasing power from the MISO wholesale market during peak demand, deferring planned retirements of existing fossil plants, accelerating renewable development, and potentially building new generation over time. The battery storage financed by the project provides some flexibility but cannot substitute for baseload generation over extended periods.

Michigan law (SB 237) requires that data centers receiving tax incentives source 90 percent of their electricity from "clean energy."[128] The clean energy requirement was the quid pro quo for the tax exemptions—a concession to environmental concerns that might otherwise have blocked the incentive package. But the definition of clean energy, the timeline for compliance, and the enforcement mechanisms remain unclear. If clean energy means zero-carbon generation, DTE's current mix falls far short: nuclear (Fermi 2) provides roughly 20 percent of DTE's generation, renewables add another 15 to 20 percent, and the balance is fossil fuel.

The law could be satisfied through REC purchases—buying certificates from wind farms elsewhere to match Saline's consumption on an annual basis. This is the approach most corporate renewable energy claims rely on, and it would satisfy the letter of the law without changing the physical electrons flowing into the data center.

The practical reality is that Saline's power will come predominantly from natural gas, at least initially. Fermi 2's nuclear output is already committed to existing customers. New renewable capacity takes years to build. The immediate source of baseload power at this scale is the same source providing marginal generation throughout MISO: natural gas turbines spinning somewhere in the region.

Over time, the picture may shift. DTE could invest heavily in renewable capacity dedicated to data center loads or pursue nuclear options, though no announcements have been made. Battery storage could help smooth demand during high-usage periods. But these changes operate on five-to-ten-year timelines. The data center begins consuming power in 2027.

The Saline project exemplifies the industry-wide pattern. Press releases emphasize clean energy commitments and closed-loop cooling systems that minimize water use. Environmental advocates note that 1.4 gigawatts of continuous demand, if powered by natural gas, produces roughly 6 million tons of carbon dioxide annually. Both perspectives capture something true, and the gap between them defines the political economy of data center development.

Harold has watched the energy debates from his kitchen window for decades. He remembers 1979, when Three Mile Island made the news and his father wondered whether the whole nuclear experiment had been a mistake. He remembers the natural gas pipelines going through in the 1990s, the wind turbines appearing on the horizon a decade later. Now heavy equipment arrives daily at his former farm. He does not know where the power will come from—nuclear, gas, wind, something not yet invented. Fifty years of farming taught him that the future has a way of surprising you.

6.8 THE CARBON FUTURE

What does data center growth mean for American carbon emissions? Projections range from concerning to alarming, depending on generation mix and technology assumptions. The estimates cited earlier in this chapter—200 million additional tons by 2030, emissions rivaling major

American cities by the mid-2030s—assume current construction patterns continue, with more than 100 gigawatts of planned capacity, much of it fossil-fueled.

These projections are not destiny. They assume current construction patterns continue and that nuclear and renewable alternatives fail to scale. If the nuclear renaissance delivers on even a fraction of its promised capacity, if battery storage costs keep declining, if renewable generation keeps growing, the trajectory could bend downward.

But we should not assume success. The history of energy transitions teaches that change is slow, expensive, and politically contested. Coal's share of US generation took thirty years to fall from 50 percent to 16 percent. Natural gas grew over decades. Nuclear expansion halted after Three Mile Island and Chernobyl. Energy systems have enormous inertia.

SMRs promise to break the pattern of nuclear delays, but promises are not power plants. And corporate sustainability commitments have a history of revision when circumstances change.

Carbon capture offers a potential middle path. Several gas-powered data center projects incorporate carbon capture systems that remove carbon dioxide from turbine exhaust and either sequester it underground or use it for industrial purposes. The Adams Fork project in West Virginia claims it will capture 99 percent of emissions. Carbon capture can reduce the carbon intensity of gas generation from 400 kilograms of CO_2 per megawatt-hour to 80-120 kilograms—still not zero, but a substantial improvement.

The economics of carbon capture depend on the 45Q federal tax credit, which provides $85 per ton of sequestered carbon.[129] At that subsidy level, capture becomes economically viable for large-scale operations. Without the credit, the technology struggles to compete, and the credit's longevity is subject to political uncertainty, creating investment risk for projects that depend on it.

The most likely scenario is a diverse portfolio: natural gas providing near-term reliability, renewables growing steadily, nuclear arriving slowly, carbon capture deployed where economics allow. This portfolio will not achieve zero emissions by 2030, but it might bend the curve downward from the most alarming projections.

6.9 THE SPEED PROBLEM

The fundamental challenge for data center sustainability is neither technology nor economics. It is time. Speed kills sustainability.

AI demand operates on six-to-eighteen-month cycles. A company training a frontier model needs power now, not in 2035. The competitive dynamics of the AI industry do not permit waiting for the perfect solution. First-mover advantage accrues to companies that deploy fastest, regardless of power source. A company that delays AI training to wait for clean power loses to competitors who train now on dirty power.

Natural gas deploys in two to three years, as described earlier. A company that needs power in 2027 can realistically achieve that timeline with gas.

Solar with storage takes three to five years. Permitting is easier than for gas, but storage systems add complexity. Battery supply chains are constrained by lithium and other materials, and construction timelines have lengthened as demand has grown.

Nuclear takes eight to twelve years or more, as the SMR timeline analysis showed. No company needing power before 2035 can rely on new nuclear capacity.

The technology that can be built fastest is the technology that gets built. Speed dominates.

This timing mismatch explains why companies simultaneously build gas plants and announce nuclear partnerships. Gas plants meet immediate needs; nuclear partnerships position for the future. Each addresses a different time horizon. The strategy is not hypocritical. It is adaptive.

The question is whether the bridge becomes permanent. Natural gas plants have twenty-to-thirty-year operating lives. Infrastructure built today will still be running in 2050. If data center growth continues at current rates, and if that growth is powered primarily by gas, the carbon commitments become impossible to meet. The bridge becomes the road.

The nuclear renaissance is a bet that the bridge will not become permanent. Technology companies are spending billions to ensure that carbon-free alternatives exist when current gas plants reach end of life. But the bet requires that SMRs deliver on their promises, that costs come down, that regulatory processes accelerate, that fuel supply chains develop. Each of these requirements is uncertain.

The worst-case scenario: nuclear delays, renewable intermittency persists, and the industry locks in decades of gas-fired generation. Carbon emissions from data centers would exceed all projections. Corporate sustainability commitments would become impossible to meet without offsets and accounting maneuvers. The AI revolution would be powered by fossil fuel.

The best-case scenario: nuclear succeeds, storage costs plummet, and the industry transitions to genuinely carbon-free operation by 2040. Emissions would still rise in the interim, but the long-term trajectory would bend toward zero. The promises would be kept—eventually.

Reality will likely fall between these extremes. Some nuclear projects will succeed; others will fail. Some regions will run largely on renewables; others will remain gas-dependent.

6.10 Looking Forward

This chapter has traced a gap: between aspiration and reality, between announcement and operation, between marketing and physics. Technology companies make aggressive sustainability commitments while building natural gas infrastructure. They announce nuclear partnerships that

157

will not deliver power for a decade. They claim renewable achievements through accounting mechanisms that do not reduce actual emissions.

The gap is not exactly hypocrisy. It reflects genuine constraints: the impossibility of waiting years for zero-carbon power when AI competition operates on months, the unavailability of sufficient renewable generation with storage to meet continuous baseload demand, the decade-long timelines of nuclear development. Press releases and sustainability reports emphasize positive developments; permit applications tell a different story, one of gas turbines and diesel backup generators. Neither document lies. They simply describe different realities.

For Saline Township, these abstractions have concrete implications. The 1.4-gigawatt facility will consume power from somewhere. If that power comes primarily from natural gas—as near-term economics and timelines suggest—the facility will generate millions of tons of carbon emissions annually. If Michigan's clean energy requirements are enforced strictly, the project may need to purchase renewable energy certificates, develop dedicated renewable capacity, or pursue nuclear options not yet announced.

The previous chapter explained why data centers are built where the grid can accommodate them. This chapter has explained what fills that grid. The next section of the book turns from the supply side—power and land—to the demand side: who wants these facilities, who pays for them, and what political bargains make them possible.

The fundamental question remains unresolved: Can the AI boom be powered sustainably, or will it become the largest new source of carbon emissions in a generation? The answer depends on whether the nuclear renaissance delivers, whether storage costs continue falling, whether carbon capture scales, whether corporate commitments translate to actual infrastructure investment. The next decade will reveal whether the promises were real or merely marketing.

Until then, the gas turbines spin.

The student in Ann Arbor has long since received an answer. The cursor has stopped blinking. The conversation has moved on to follow-up questions, clarifications, the easy back-and-forth that feels like talking to a very patient tutor. The coffee has gone cold.

What the student does not see: somewhere in the regional grid, a natural gas turbine adjusted its output to accommodate the load. The fuel that spun that turbine was extracted from shale formations deep beneath Pennsylvania or Ohio, processed through compressor stations, and piped hundreds of miles to a power plant that converts chemical energy into electrical current. The carbon released by that combustion is now dispersing into the atmosphere, joining the 36 billion tons humanity adds each year.

We have traced the token's journey through the full physical stack. From the mathematics of attention mechanisms to the silicon pathways of GPUs. From the liquid-cooled racks to the buildings housing them. From the transmission lines carrying power to the generators producing it—gas turbines spinning, nuclear reactors sustaining controlled fission, wind turbines turning in Kansas fields, solar panels absorbing desert sunlight.

The electricity is real. The carbon is real. The water evaporating from cooling towers is real. Every query has a physical cost, distributed across infrastructure that spans continents.

But infrastructure requires land. The Saline Township campus sits on 575 acres that grew corn and soybeans a generation ago. How did farmland become the destination for a question asked in a coffee shop? The answer involves economics that most people never consider—the strange logic that makes rural acreage more valuable for computation than the urban centers where the questions originate.

CHAPTER SEVEN

The Land

T HE REALTOR'S SIGN STILL STANDS at the edge of the driveway, but it no longer matters. Three generations of *Harold*'s family farmed this land in Saline Township, growing corn and soybeans on five hundred acres of Kibbie fine sandy loam and Pewamo clay—soil names his grandfather knew by feel before any county survey existed. Now the land belongs to Related Digital, purchased for a sum that *Harold*, the last farmer to work it, describes as "more money than I ever imagined seeing in my lifetime."

Harold is seventy-three years old. His hands show it: thick knuckles, skin like cracked leather, a scar across the right palm where a combine auger caught him in 1987. His grandfather bought this farm in 1922, clearing the woodlot with a team of horses and building the farmhouse that *Harold* was born in and lived in for seven decades. His father expanded through the Depression, buying foreclosed neighbors' land at prices that haunted him until he died. *Harold* himself survived the farm crisis of the 1980s. He remembers the auction signs, the silence where tractors used to run.

He raised two children here, both of whom moved to cities and never looked back. His daughter teaches in Chicago. His son writes software in Seattle. Neither wanted the farm. Neither wanted the early mornings, the equipment breakdowns, the weather anxiety, the margins so thin that one bad year could wipe out three good ones. He understood. He had wanted them to have easier lives than his.

"They weren't going to farm it," *Harold* says, standing at the edge of his former property. The wind smells like diesel now, not soil. "And the offers kept getting bigger. First one price, then double, then triple. I held out as long as made sense." He pauses, looks at the ground his grandfather cleared. "Maybe longer."

He thinks about what the surveyors said when they walked the property—nice flat site, easy access, good drainage. They did not understand what they were looking at. The county soil survey mapped his fields as Kibbie fine sandy loam on the high ground, Pewamo clay loam in the low spots, Morley loam on the gentle slopes between—classifications that meant nothing to the surveyors but everything to *Harold*. His grandfather had learned which soil held water and which drained fast, which warmed early in spring and which stayed cold. Three generations of cover crops and manure and crop rotation had built six inches of topsoil where there used to be two. His grandfather had started with glacial till, rocky and acidic. His father spent decades working in lime and organic matter. *Harold* himself had adopted no-till practices in the nineties, letting the earthworms do what plows could not. The soil they were about to bury under concrete had taken a century to create. It would take a century to rebuild somewhere else, if anyone bothered. Probably no one would bother.

What he does not say: the night he signed the papers, he could not sleep. He walked the fields at 3 a.m., frost crunching under his boots, trying to memorize every contour. The rise where his father taught him to drive a tractor. The low spot that flooded every spring and grew the sweetest corn. The oak at the property line where he carved his initials as a boy and later his children's initials and later still, nothing, because there was no one left to carve.

Seven and a half million dollars. More money than three generations of his family had earned combined. Enough to pay off the equipment loans, set up trust funds for grandchildren he rarely saw, buy a condo somewhere warm where his wife's arthritis would not have ached every winter. Enough to make the decision rational, if any amount could.

He still wakes some nights thinking he hears the combine running.

Not every farm sold. A few blocks northeast of downtown Saline, H&K Braun Farms still operates—a Michigan Centennial Farm, founded by George Braun in 1907, the metal plaque by the road donated by DTE Energy certifying one hundred years of continuous family ownership. Howard Braun runs the operation now with his nephew Karl, raising corn, soybeans, and wheat on the same Sisson fine sandy loam where the family ran dairy cattle for ninety years until 1998. Howard's brother Kelven worked beside him until his death in 2017 at eighty-two. Four generations on the same soil, watching the town change around them: the interstate that bypassed the old Chicago road, the subdivisions that swallowed neighboring farms, and now the data centers rising to the west. The DTE plaque on their fence post carries a quiet irony—the same utility now supplying a gigawatt to facilities that arrived after their family had been farming for a century.

The price per acre that Related Digital paid for this farmland—around fifteen thousand dollars—sounds extravagant for agricultural property. Comparable farmland in Washtenaw County typically sells for eight to ten thousand dollars an acre. But for a company planning to invest seven billion dollars in a data center campus, the land acquisition represents a rounding error. The entire five hundred acres cost less than one percent of the project budget.

This is the story of land in the data center era. Agricultural families sell property for record premiums. A few miles away, abandoned factories and brownfield sites sit unused. Brownfields are former industrial properties, often contaminated or complicated by prior use. The paradox frustrates policymakers, environmentalists, and community advocates alike. It also represents one of the most significant transformations of American land use since the postwar suburban expansion.

The farmland paradox reveals how technical decisions embed political choices. The political theorist Langdon Winner made this point in a famous 1980 essay: artifacts have politics.[130] The technologies we build, the infrastructure we create, the sites we choose—these are not neutral. They embed power relationships that persist long after the choices are made. Winner's most memorable example was Robert Moses, the urban planner who designed Long Island's parkway overpasses too low for buses to pass under. The choice seemed purely technical, a matter of bridge engineering. Its effect was political: Black and poor New Yorkers who relied on public transit could not reach the public beaches. The concrete encoded the exclusion.

Data center siting works the same way. The "technical" preference for greenfield sites—undeveloped land, typically agricultural—over brownfield concentrates costs on rural communities with minimal political power while directing benefits to corporate shareholders and urban consumers. The choice of where to build a data center is a political choice, even when it presents itself as pure engineering.

7.1 THE FARMLAND PARADOX

America has thousands of abandoned industrial sites—steel mills, coal plants, refineries—sitting idle while data centers consume hundreds of thousands of acres of productive farmland. Policy goals favor brownfield reuse. Environmental advocates push for it. Local communities want their abandoned sites revitalized. Yet the market delivers the opposite.

The explanation begins with power infrastructure, but extends far beyond electricity. Data centers occupy farmland because farmland is cheap, clean, fast, and simple. Brownfield sites are expensive, contaminated, slow, and complicated. No policy preference has proven sufficient to overcome these advantages.

In 2025, I documented over six hundred data center projects in the United States representing more than one trillion dollars in investment.[8] The pattern holds regardless of region, sponsor, or project size. Microsoft chooses farmland. Blackstone chooses farmland. CoreWeave chooses farmland. The market speaks with one voice.

Consider the geography of recent announcements. Project Kestrel targets nearly four hundred acres of farmland north of Kansas City. The Related Digital campus in Saline occupies former corn and soybean fields. Meta's proposed Howell Township facility would have consumed over a thousand acres of agricultural property. From Virginia to Texas, the pattern repeats. Data centers are devouring farmland.

This defies the intuitions of urban planners and environmental advocates. Abandoned industrial sites offer apparent advantages: existing roads, some utility connections, communities eager for revival. Many sit in areas that lost manufacturing jobs decades ago and would welcome new investment. Yet developers consistently pass them over for pristine agricultural land.

To understand why, we need to examine the economics, timelines, and risks driving site selection. These factors create what the industry calls the "farmland paradox"—the persistent preference for pristine agricultural land over apparently available industrial sites.

7.2 THE ECONOMICS OF LAND

The cost differential between farmland and industrial brownfield sites is staggering. Agricultural land in the key data center states sells for two to ten thousand dollars per acre.[131] Industrial brownfield in the same regions runs fifty to two hundred thousand. Urban industrial land in metro areas can exceed one million dollars per acre.

For a gigawatt-scale data center requiring five hundred to one thousand acres, this differential means paying five million dollars for land—or five hundred million. A hundred-fold difference. Yet for a $10 billion project, even the most expensive land represents only 5 percent of total budget. Five hundred acres of Midwest agricultural land costs $2.5 to $5 million; industrial brownfield costs $25 to $100 million; urban industrial land runs $250 to $500 million. At data center scale, land cost is the least important factor in site selection.

Consider the Related Digital project in Saline Township. The land cost of approximately eight to nine million dollars covered multiple parcels totaling 575 acres; *Harold*'s five hundred acres accounted for the bulk of that figure. Even at those prices, land represents roughly 0.1 percent of the seven billion dollar budget. Had the company paid ten times more for brownfield land, the increase would still represent only about one percent of total project cost. At data center scale, land is essentially free.

The same mathematics apply everywhere. Project Kestrel's roughly three million dollar land acquisition represents 0.003 percent of its hundred billion dollar budget. Meta's billion-dollar-plus Howell Township proposal would have involved land costing perhaps ten million. AWS's twenty billion dollar Falls Township development includes land acquisition costs that barely register.

So why do developers care so much about land type if price barely matters? The answer lies in hidden costs and timeline impacts. Land price is a rounding error. Land complexity can kill a project.

7.3 THE THREE BROWNFIELD BARRIERS

Every brownfield site carries three categories of risk that greenfield sites avoid: contamination, infrastructure gaps, and political complexity. Any one can add years and hundreds of millions of dollars to a project. Together, they make brownfield development economically irrational for most data center applications.

Former industrial sites carry environmental legacies that must be addressed before construction can begin. The contamination varies by prior use: heavy metals from metalworking, petroleum hydrocarbons from manufacturing, PCBs from electrical equipment, asbestos in structures, coal ash from power plants. Each type demands specific remediation approaches with different costs and timelines.

The assessment process alone consumes time and money. A Phase I Environmental Site Assessment—reviewing historical records, identifying contamination sources—costs two to six thousand dollars and takes two to four weeks. Phase II goes deeper: actual soil and groundwater sampling, five to one hundred thousand dollars, four weeks to several months. If contamination turns up, remediation begins. Here costs and timelines become anyone's guess.

Soil remediation costs fifty to five hundred dollars per cubic yard. A major industrial site might require removing hundreds of thousands of cubic yards. Groundwater treatment runs five hundred thousand to five million dollars per year and can continue for decades. Demolition with hazardous material handling adds ten to fifty million dollars for major facilities. Vapor barriers, monitoring wells, and treatment systems pile on further costs.

The remediation process introduces its own regulatory complexity. Environmental agencies at federal, state, and sometimes local levels must approve remediation plans, certify that cleanup goals have been achieved, and grant the deed restrictions or environmental covenants that allow development to proceed. Each touchpoint introduces delay risk.

Some industrial sites appear clean on initial assessment but reveal contamination during excavation. A 2008 study found median brownfield remediation costs of fifty-seven thousand dollars per acre—but this figure masks enormous variation.[132] Sites with significant contamination can cost millions per acre to remediate. The discovery of unexpected contamination mid-construction creates the worst scenario: a project halted indefinitely with money already spent and no clear path forward. Construction crews encountering contaminated soil must stop work, notify regulators, conduct additional assessment, and potentially redesign foundations and drainage systems. These surprises can add one to two years to project timelines.

The total remediation premium for a major brownfield site ranges from fifty to two hundred million dollars. This cost alone would not disqualify brownfield development—it remains a small fraction of a multi-billion dollar project. But remediation introduces schedule risk that no amount of money can fully mitigate.

Unknown contamination discovered during excavation can delay projects indefinitely. Construction loans carry interest. Hyperscaler customers expect delivery dates. Investors demand returns on schedule. Every month of delay bleeds money—and in the AI infrastructure race, where first-mover advantage is measured in months, schedule uncertainty matters more than cost.

The Nautilus Data Technologies project at Stockton Port in California illustrates this timeline challenge. Despite designing a water-cooled facility with minimal environmental impact, the company spent three and a half years navigating regulatory approval for its brownfield site. The biggest risk to brownfield redevelopment, as one industry executive put it, is "all the projects that never get built—that die on the vine" during extended approval.

A common assumption holds that brownfield sites should have infrastructure advantages: existing roads, utilities, and power connections from prior industrial use. This assumption proves largely false for data centers.

Former industrial sites were designed for twentieth-century manufacturing loads, not twenty-first-century computing demands. A typical
factory required five to twenty megawatts of power. A hyperscaler data
center requires five hundred megawatts to one gigawatt or more. The existing electrical infrastructure cannot be "upgraded"—it must be rebuilt
entirely.

Consider what a gigawatt-scale data center actually needs. Multiple
redundant transmission feeds at 345 kilovolts or higher. Dedicated substations with transformer capacity measured in hundreds of MVA. Distribution systems to power dozens of buildings. Backup generators with
fuel for days. None of this exists at former factories.

Worse, the interconnection queue position—the place in line for utility connection, managed by regional grid operators—that determined
the timeline for the original industrial connection has long since lapsed.
When a developer requests gigawatt-scale power to a brownfield site,
they enter the same utility queue as any other new customer. The existing infrastructure provides no advantage.

In some cases, brownfield sites face longer interconnection timelines
than greenfield alternatives. Legacy infrastructure must be decommissioned before new equipment can be installed. Legacy agreements with
multiple utilities may need resolution. Right-of-way questions may require renegotiation. Existing underground utilities—gas, water, sewer,
communications—may need relocation. The net result: brownfield transmission timelines of three to five years versus two to three for greenfield.

This reflects a deeper irony. Data centers seek locations along high-
voltage transmission corridors precisely because these corridors enable
rapid interconnection. But transmission corridors run through open
land—farmland, not former factories. Twentieth-century industrial sites
clustered near labor pools and transportation networks. Twenty-first-
century data centers cluster near electrical substations and transmission
lines. The geographic logic has shifted.

Water infrastructure presents similar challenges. Former industrial sites may have water connections, but not at the volumes data centers require. A gigawatt-scale facility with evaporative cooling needs three to five million gallons per day—comparable to a small city. Existing water mains sized for factory use cannot deliver this volume. New mains must be installed, often requiring easements across multiple properties.

Fiber connectivity follows its own geographic logic. Major fiber routes run along highways and railroad rights-of-way, which may or may not pass near brownfield sites. Installing fiber to a remote brownfield site can cost millions and take months.

———

Brownfield sites accumulate political complexity over decades of industrial use, decline, and abandonment. Developers seeking rapid execution face obstacles that simply do not exist on greenfield sites.

Title complexity comes first. Agricultural land typically belongs to one family, maybe two, with straightforward deed histories. Brownfield sites are different: multiple corporate successors from bankruptcies, unresolved environmental liability chains, tax liens from multiple jurisdictions, easement complications, union pension claims, EPA cost recovery liens. The title search alone can take months.

Consider a typical former steel mill. The original company may have been acquired multiple times, each transaction creating successor liability questions. Environmental enforcement actions may have created federal liens. State and local tax delinquencies may encumber the property. Former employees may hold retirement benefit claims against corporate assets. Bankruptcy proceedings may have left competing claims unresolved.

Clearing title on a complex brownfield site can take one to three years of legal work. Multiple law firms may be required to address different claim types. Settlement negotiations with governmental and private claimants can extend indefinitely. Meanwhile, the developer cannot proceed because they cannot establish clear ownership.

Clearing title on farmland takes two to four weeks.

Community expectations create the second barrier. Former industrial communities often see their abandoned sites as opportunities for comprehensive redevelopment—mixed-use projects bringing retail, housing, diverse employment. A data center provides tax revenue and some jobs, but does not rebuild a community. Residents who expected revitalization may oppose a project that delivers only a massive building with minimal permanent staff.

This expectation gap creates political risk. Community leaders who initially welcomed data center interest may face backlash when the project's limited employment becomes clear. Local officials may impose additional requirements—community benefit agreements, local hiring mandates, affordable housing contributions—that add cost and complexity. Public hearings grow contentious as residents voice disappointment.

Regulatory overlay adds the third barrier. Brownfield sites may fall under multiple environmental enforcement regimes. CERCLA—the Comprehensive Environmental Response, Compensation, and Liability Act, better known as Superfund—can make present and prior owners jointly and severally responsible for full cleanup costs. State environmental agencies may impose separate requirements. Local jurisdictions may raise historic preservation concerns. Each touchpoint creates delay risk and potential veto points.

Pennsylvania coal country illustrates this expectation gap. Towns devastated by coal plant closures hoped data centers would choose their sites, bringing jobs and tax revenue to replace what mining took away. Instead, most Pennsylvania data center investment has gone to greenfield sites or relatively clean brownfield locations like former printing plants. The coal communities remain waiting, their contaminated sites too costly and complex for rapid development. The jobs that left with the mines have not returned.

The combined effect of these three barriers—contamination, infrastructure, and complexity—creates a consistent timeline gap. Greenfield sites break ground in six to twelve months from acquisition. Brownfield sites require two to four years. In a market where demand doubles annually and competitors race to capture hyperscaler contracts, two to three extra years represents an insurmountable handicap.

These barriers apply at scale. Smaller facilities face different math. The Deep Green project in Lansing requires no remediation. The power demand fits existing grid capacity. The approval process runs through City Council, not lawsuit and settlement. At 24 megawatts, the brownfield barriers that make gigawatt-scale urban development impossible simply do not apply. But 24 megawatts cannot serve a hyperscaler training the next generation of AI models. The scale that creates the demand also creates the constraints that push development onto farmland.

Security requirements add a fourth consideration that receives less attention but matters significantly. Data centers require defensible perimeters: clear sight lines, controlled access points, space for security patrols. Rural farmland offers all of this naturally. A flat field surrounded by other flat fields provides excellent visibility. Fencing the perimeter is straightforward. There are no adjacent buildings from which someone might observe operations, no shared walls, no complex property boundaries creating blind spots or access disputes.

Urban and brownfield sites present the opposite. Adjacent properties mean shared boundaries. Multi-story neighbors provide elevated vantage points. Complex site geometries create security dead zones. The hyperscalers running AI infrastructure treat their facilities as high-security installations—not because they fear physical theft, but because they protect proprietary model weights, customer data, and competitive intelligence. A cornfield offers better security than an urban infill site, and security teams factor this into site selection recommendations.

7.4 WHY POLICY HASN'T WORKED

Given the policy preference for brownfield redevelopment, we might expect state and federal programs to offer incentives sufficient to overcome these disadvantages. Such programs exist, but they have proven inadequate to shift developer behavior.

The federal brownfield tax deduction—which allowed remediation costs to be deducted in the year incurred rather than capitalized—expired in January 2012 and has not been renewed despite multiple legislative attempts.[133] Congress has considered brownfield incentive restoration in various tax packages, but the provision consistently loses to other priorities. The Inflation Reduction Act's Energy Community Bonus provides an additional ten percent tax credit for clean energy projects on brownfield or coal community sites, but this bonus applies to power generation, not data centers.[134]

EPA's Brownfields Program provides grants and technical assistance for assessment and cleanup, but funding remains modest relative to remediation costs. A typical grant of two hundred thousand dollars barely covers Phase II assessment for a major site, let alone actual remediation. The program has helped smaller redevelopment projects but cannot meaningfully influence billion-dollar data center decisions.

State incentive programs rarely distinguish between greenfield and brownfield sites. Virginia's hundred percent sales tax exemption applies regardless of site type. Texas property tax abatements work the same on former farmland as on former refineries. Ohio's data center exemptions give no preference to brownfield locations. When incentives are equal, developers choose the lowest-cost, fastest path. That path runs through cornfields.

Michigan attempted the most innovative approach in January 2025, creating differentiated incentives: tax exemption through 2050 for standard sites, but extension through 2065 for brownfield or former power plant sites. This fifteen-year differential aimed to compensate for brownfield timeline penalties with extended tax benefits. The logic was sound: if brownfield sites cost two to three years of delay, extend the benefit period to offset those costs.

The Michigan experiment lasted eleven months. By December 2025, bipartisan legislation sought to repeal or significantly modify the program. Republicans objected to the fiscal cost of long-duration exemptions. Democrats complained that brownfield uptake remained minimal while greenfield projects captured most benefits. Industry representatives found compliance requirements burdensome. Environmental groups noted the program had not actually redirected development to brownfield sites.

The most serious attempt to redirect data center siting faced elimination before it could demonstrate results. The political reality proved what economic analysis predicted: incentives strong enough to change developer behavior impose fiscal costs that elected officials find difficult to sustain. Developers cannot rely on programs that may be repealed before projects complete.

7.5 Case Studies in Land Selection

The farmland paradox manifests differently across regions, but the underlying economics hold. Four case studies illustrate the pattern. Kansas farmland shows pure greenfield success. Pennsylvania brownfield yields mixed results. Arizona desert reveals greenfield with emerging resistance. Michigan marks the frontier of community opposition.

Project Kestrel, the hundred billion dollar development announced for Kansas in 2025, represents the greenfield model in its purest form.[135] The project acquired 379 acres of agricultural land at prices estimated around seventy-five hundred dollars per acre—a total land cost of approximately three million dollars, a fraction of a percent of budget.

Kansas offered minimal brownfield inventory and even less policy pressure to consider it. The state won on speed: utility Evergy committed to power on an aggressive timeline, officials streamlined permitting, and landowners sold at prices that seemed generous by farming standards but trivial at project scale.

From announcement to construction start, Project Kestrel expects to move in under eighteen months. Environmental assessment requires three to four months for standard agricultural review. Title clearance takes two to four weeks. No contamination assessment, no remediation, no complex title chains. The farmers who sold received windfalls—but the land was a rounding error, costing less than a single data hall to build.

Pennsylvania provides the most comprehensive test of brownfield viability, with multiple projects attempting different approaches.

TECfusions Keystone Connect targets the former Alcoa R&D campus near Pittsburgh for a twenty billion dollar, three gigawatt development. The site offers relatively favorable brownfield conditions: industrial R&D rather than heavy manufacturing. Yet the timeline has extended beyond expectations as environmental assessment and infrastructure modernization proceed.

AWS Falls Township pursues the former U.S. Steel mill site for twenty billion dollars.[136] The Fairless Works produced steel from 1952 until 2001, leaving slag piles, contaminated groundwater, and asbestos-containing structures. AWS chose this location partly because it had already committed to Pennsylvania through a nuclear co-location project. Political

175

pressure encouraged brownfield demonstration. But even AWS, with un-limited capital and patient timelines, faces years of site preparation before construction can begin.

CoreWeave Lancaster selected former printing plants for six billion dollars. Printing represents clean industrial use—no heavy metals, no hazardous waste legacy. The site succeeds precisely because its prior use was light industrial. Light industrial can work. Heavy industrial rarely does.

Homer City (Chapter 6) illustrates the power plant paradox. The ten-billion-dollar project targets a former coal plant with exceptional grid infrastructure from decades of power generation—but coal ash requires extensive remediation. The best grid connections often come with the worst environmental legacy.

Pennsylvania's experience suggests brownfield development works when prior industrial use was relatively clean, when grid infrastructure provides significant advantage, or when developers have parallel green-field projects to meet timeline demands elsewhere.

Arizona demonstrates that greenfield development is not frictionless. Water constraints and community opposition can kill projects even on undeveloped desert land.

Vermaland La Osa Data Center Park proposes thirty-three billion dollars on 3,300 acres of desert near Eloy, south of Phoenix. Classic green-field: remote location, minimal existing development, state-level politi-cal support. But Arizona's water challenges are intensifying. The state has long overdrafted its aquifers. Critics question whether groundwater can sustain millions of gallons of daily consumption. Agricultural users worry about competition for already-scarce water.

AWS Project Blue in Tucson was canceled entirely due to local op-position over water consumption. Residents organized; public hearings drew crowded resistance. A hyperscaler with a multi-billion dollar budget walked away because community opposition proved too strong.

Arizona's trajectory may shift from data-center-friendly to data-center-cautious as water constraints intensify. Some communities are considering data-center-specific water restrictions or outright moratoriums. The greenfield advantage depends on political acceptance—and that acceptance cannot be assumed indefinitely.

Michigan in 2025 became the leading edge of community opposition to data center development. Multiple projects faced organized resistance, ballot initiatives, and outright denial.

In Howell Township, a Meta-backed proposal for a hyperscale data center on over a thousand acres met fierce resistance. The Planning Commission voted unanimously to recommend denial after a seven-hour meeting with hundreds of opposing comments.[137] The developer withdrew the rezoning request before the township board could vote, though the township adopted a six-month moratorium—and residents expect the project or something like it to return.[138]

Augusta Township approved rezoning for a Form8tion/Thor Equities billion-dollar data center, only to face a citizen referendum. The grassroots group PACT collected 957 signatures—nearly double the 561 required—to force a 2026 ballot vote.[139]

Kalkaska County saw a proposed hyperscale data center on 1,440 acres of state forest land withdrawn after public opposition.[140] Critics focused on using public forest for private commercial development.

Even Ypsilanti Township, which hosts the proposed University of Michigan / Los Alamos National Laboratory data center, has seen organized resistance.[141] This project is backed by a public university and a national laboratory, focused on research rather than commercial AI. It still faces "significant local opposition."

The Michigan experience shows that community opposition has reached new intensity. Projects that would have sailed through five years ago now face organized resistance. Whether this pattern spreads to other states will shape where the industry can build.

7.6 THE SALINE TOWNSHIP DECISION

Frank's phone buzzed at 6:15 AM on a Saturday—the third call that week from a number he did not recognize. The Related Digital project had been approved, the consent agreement signed, the lawsuits settled. But the calls kept coming. "You sold us out," one voicemail said. "Three generations of farmers, and you let them pave it over for computers."

He listened to the voicemail in his kitchen, coffee growing cold, while his wife pretended not to hear. Forty years of marriage. She knew when to give him space.

Frank understood the anger better than the callers knew. His own father had farmed eighty acres until 1974, when the costs finally overtook the revenue. He knew what farming did to families.

He had grown up in Saline Township, watched the farms that fed his family give way to subdivisions, then strip malls, then the quiet decay that came when young people moved to cities and old people died. His son moved to California years ago, but his daughter *Beth* stayed closer— she works in Troy now, in mortgage processing for a regional bank. She is the one who brings *Lily* and *Marcus* for Christmas, who calls on Sundays, who worries about her father the way he used to worry about his.

His wife mentioned that *Harold* had called again, sounding lost. "He wanted to know if you thought he did the right thing," she said. "I told him you weren't home."

Frank remembered their conversation last fall, before *Harold* signed. "Whatever you decide, I understand." That was what he had said. He had meant it. He still meant it. But understanding was not the same as knowing what to say now.

He thought about the decisions that had led to this moment. The zoning vote, when he had raised his hand almost without thinking. The lawsuit, which arrived two days later like a punch to the gut. The consent agreement with its provisions for groundwater monitoring and well mitigation, negotiated in conference rooms where the lawyers did most of the talking. He had believed those provisions would protect the community. Now he was not sure protection was possible.

He thought about calling *Harold* back. What would he say? That the money was good and the decision was rational and the township could not have won anyway? That sometimes you lose even when you do everything right?

He did not call.

He met *Ellen* for coffee sometimes, at City Limits Diner on Michigan Avenue—a greasy spoon where the menu runs from goulash to gluten-free and the waitresses know your name. They sat in the same booth each time, watching the trucks go by, talking about nothing and everything. She never asked him about the vote, never asked if the consent agreement had been the best they could do. He was grateful for that.

Why did Saline Township attract Related Digital's investment? The answer illustrates every factor we have examined: power availability, land economics, timeline requirements, and political context.

The transmission corridors connecting DTE Energy's generation to southeastern Michigan run through Saline Township—the same infrastructure that made *Harold*'s land valuable. DTE has agreed to supply 1.4 gigawatts.[142] The utility's willingness reflects its need for reliable baseload demand to justify infrastructure investment as traditional loads decline.

The land was agricultural—clean, simple, available. *Harold* and his neighbors sold at premiums that seemed generous by farming standards but trivial by data center economics. Title clearance took weeks. Environmental assessment was routine. No contamination lurked beneath the soil.

The site offered other advantages: proximity to major highways for construction logistics, adequate water from regional systems, and fiber connectivity along existing routes. These factors matter less than power availability but contribute to the site's appeal.

The township offered political acceptance that other Michigan communities refused. In Howell Township, a Meta-backed proposal was withdrawn after the planning commission recommended denial. Augusta Township faces a citizen referendum. Saline's path—initial denial, lawsuit, consent agreement—was described in the prologue. What distinguishes the outcome is that litigation pressure produced negotiated settlement rather than impasse, with provisions for groundwater monitoring and well mitigation that neither side would have accepted voluntarily.[56]

Ellen still lives on the farm adjacent to the project site—eighty acres of Boyer loamy sand and Spinks loamy sand, the farm where her grandfather dug the well that has served her family for three generations. Four months after the board meeting where her voice cracked reading from handwritten notes, the first quarterly monitoring report arrives. No adverse impacts detected. Water table stable. Contaminant levels within normal parameters.

She reads the report at her kitchen table, the same oak table where her mother served breakfast for forty years, its surface scarred by a thousand meals and marked with a faint ring where her father's coffee cup always sat. The morning light comes through the window at the same angle it always has, catching the dust motes, illuminating the philodendron on the sill that her mother planted in 1978 and that *Ellen* has somehow kept alive ever since. The report is twelve pages of technical language and charts. She puts on her reading glasses—the same frames she wore to grade biology exams for three decades—and works through it slowly. She understands enough to know what matters: the well is fine. For now.

But through her kitchen window, where she used to watch deer cross the back field at dawn, she sees the construction site. *Harold's* former farm, where she caught fireflies as a child, where she walked with her husband on summer evenings before his heart gave out, is now a sea of excavators and concrete forms. Cranes rise against the morning sky like strange mechanical birds. The sound of diesel engines has replaced the sound of wind through corn. Some mornings she forgets, just for a moment when she first wakes up, and then she looks out the window and remembers.

Her sons called last week, both of them, a rare coordination that suggested they had discussed her beforehand. The older one works at the university in something called "computational biology." The younger started a company in Ann Arbor that does something with artificial intelligence. Neither of them understood why she would not sell. Both thought she should consider it while land prices were high. She had listened, said she would think about it, hung up the phone and sat at the kitchen table for an hour without moving. The philodendron needed watering. She watered it.

"It's not about the money," she told them. They did not understand. They had grown up in this house, run through these fields, drunk water from that well. But they had also left. They had made lives elsewhere. The farm was a place they visited, not a place they lived.

Last month, reviewing her quarterly retirement statement, she noticed something she had never paid attention to before. The TIAA account she had contributed to for thirty years of teaching biology listed its largest holdings. Microsoft. NVIDIA. Alphabet. The same companies building on *Harold's* land. Her retirement had grown twenty-three percent in 2024, more than any year she could remember. The same companies her younger son's business depends on were making her wealthier while they paved over the fields she had walked since childhood.

She thought about *Linda*. *Linda* taught chemistry in the room next door for twenty-two years and had planned to retire in June 2009. Then the market crashed. *Linda*'s account lost forty percent of its value in four months. She stayed five more years, complaining about her knees and the new freshmen who couldn't balance equations, until she finally retired at sixty-seven. Eighteen months later, a stroke. She never got to see the Grand Canyon. Never got to visit her grandchildren in Seattle for more than a week at a time. Thirty years of teaching, and the market decided she didn't get to rest.

Ellen understood the irony of her own gains. She understood something else, too. She had been twenty-six in 1985, the year the farm crisis hit bottom. She remembered the auction signs. The *Hendersons*, three miles down the road, who had borrowed against their land when corn was three dollars a bushel and lost everything when it fell below two. The bankers had been happy to lend when prices were rising—happy to take their fees, happy to tell everyone that land was the safest investment in the world. Then prices fell and those same bankers took the land. Mr. *Henderson* drove his truck into the pond behind the barn. Nobody called it suicide. Everyone knew.

That was the thing about money people. They made their cut on the way up and they made their cut on the way down, and the people left holding the land or the debt or the worthless shares were always somebody else. It had been true in 1985. It had been true in 2008. It would be true the next time, whenever the next time came.

The numbers on her statement could vanish the same way *Linda*'s had. But *Harold*'s land would still be gone. The concrete would still be poured. The cranes would still stand where the corn used to grow. If the bubble burst, she would lose both: the paper wealth and the land. The companies would write off their losses and find the next opportunity. The farmland would not come back.

"The water is fine," she says now, to the empty kitchen. The report sits on the table. Outside, the cranes keep moving.

She wonders what her grandfather would think. The man who dug fifty-two feet through clay and sand, who trusted that the water would always be there, who never imagined a world where someone would build a building the size of a small city next to his well.

"For now," she adds. No one hears.

The Saline outcome reveals how community acceptance has become a variable rather than a constant. Opposition is growing across Michigan and nationwide. Environmental concerns about water consumption, carbon emissions, and land conversion are crystallizing into organized resistance. Some communities will continue to welcome data center investment. Others will not. The companies that succeed will be those that navigate community relations as carefully as they navigate utility interconnection.

7.7 WATER: THE EMERGING CONSTRAINT

Land is one resource constraint. Water is another—and it is becoming more significant as data center scale increases.

Data centers require water for cooling. Traditional air-cooled systems use less water but consume more electricity to achieve the same cooling effect. Evaporative systems use water directly, achieving higher efficiency but consuming significant volumes. Liquid cooling systems, increasingly necessary for high-density AI workloads, may use closed-loop designs that minimize water consumption or open-loop systems that require makeup water.

The physics are straightforward. Computing generates heat. Heat must be removed to prevent equipment damage. Moving heat requires either air, which is inefficient at high densities, or water, which is efficient but consumptive. The choice between cooling methods involves tradeoffs among water use, energy use, and capital cost.

A traditional hundred-megawatt data center might consume half a million gallons of water per day using evaporative cooling. A gigawatt-scale facility could consume five million gallons daily or more, depending on cooling technology and climate. In water-rich regions, this is manageable. In water-stressed regions, it creates conflict with agricultural, residential, and environmental users.

Cooling technologies vary in water intensity. Air-cooled systems use little water but consume 10 to 15 percent more energy. Evaporative cooling is most efficient but consumes 0.5 to 1.5 gallons per kilowatt-hour. Closed-loop liquid systems minimize makeup water at high capital cost; open-loop liquid systems use moderate water for specialized applications. Hybrid systems adjust based on conditions. At the extremes, a one-gigawatt facility with evaporative cooling may consume three to five million gallons per day.

In aggregate, data center water consumption is modest. Lawrence Berkeley National Laboratory estimates US data centers consumed approximately 17 billion gallons for cooling in 2023.[61] American golf courses consume approximately 760 billion gallons annually.[143] Against such benchmarks, data centers appear insignificant.

Industry advocates deploy this comparison defensively. But it is misleading: golf courses are dispersed across sixteen thousand locations nationwide, while data centers concentrate in specific clusters. Google's thirstiest data center, in Iowa, consumes 2.7 million gallons daily—equivalent to water use by roughly 35,000 residents.[144]

Geography compounds the concentration problem. Northern Virginia's data centers increased county water consumption by more than 250 percent between 2019 and 2023, with usage peaking during drought-prone summer months.[145] Arizona attracts data center investment for its cheap land—and faces the Southwest's worst drought in twelve hundred years. Golf course water use has declined 29 percent in recent years; data center water use is projected to double or quadruple by 2028.[61]

Data center water consumption is not a national crisis. But it is a localized problem that creates genuine conflicts in specific communities. The solution lies in regional water planning, mandatory disclosure, and efficiency standards—not dismissal of legitimate local concerns.

Arizona's water constraints have already forced project cancellations. Other water-stressed regions—Nevada, New Mexico, parts of Texas—may follow. The industry is responding with investments in dry cooling and closed-loop liquid systems, but these add cost and complexity. The evaporative cooling described earlier remains the lowest-cost option where water is available.

In Arizona, farmers like *Roberto*—whose family has worked the same land for three generations—watch their irrigation allocations shrink while data center permits are approved. The trade-off between agricultural water and industrial cooling is not abstract. It is measured in fallowed fields and failed crops, in family farms that may not survive another decade. The water that once grew alfalfa now evaporates from cooling towers. The land remains, but its purpose has changed.

Microsoft has committed to becoming "water positive" by 2030, meaning it will replenish more water than its data centers consume.[146] The commitment involves investments in water restoration, watershed protection, and efficient cooling technologies. Google and other hyperscalers have made similar pledges. These commitments acknowledge that water consumption has become a material concern.

Water constraints interact with land selection in complex ways. Brownfield sites in water-rich regions may become more attractive if water-stressed greenfield regions become politically closed. Michigan, with its Great Lakes resources—the largest surface freshwater system on Earth—might seem positioned to benefit. But the region's history is littered with extraction regrets: Chicago's canal diversions, Nestlé's bottled water battles, agricultural drawdowns that lowered lake levels for years. Communities fought for decades to protect these waters. They may not welcome a new category of industrial demand.

But Great Lakes water is neither unlimited nor uncontested. The Great Lakes Compact restricts diversions outside the basin.[147] Withdrawals within require permits and face restrictions. Data centers seeking Great Lakes water must navigate a regulatory framework designed to protect the lakes for future generations.

7.8 COMMUNITY IMPACT

What happens when a gigawatt-scale data center arrives in a township of 2,300 people? The question matters because data center siting increasingly concentrates in rural areas where the impact is most significant relative to existing activity.

The economic impact is substantial. A multi-billion dollar construction project brings thousands of temporary jobs. Peak construction employment for a gigawatt-scale facility can exceed five thousand workers. These workers need housing, food, transportation, and services. Local businesses see demand surge. Hotels fill. Restaurants add staff. Convenience stores stock up.

Permanent operations require hundreds of workers—fewer than traditional manufacturing but more than many rural employers offer. Data center technicians, electricians, security personnel, and facilities managers earn wages well above regional averages, with health insurance and retirement plans that local businesses struggle to match. For workers who land these jobs, the opportunity is real—stable employment, good benefits, a path to middle-class security. For local employers who lose their best electricians and mechanics to competitors paying fifty percent more, the calculus is harder. The auto shop that trained a technician for five years watches him leave for the data center. The farm that relied on a skilled equipment operator finds she can earn more maintaining backup generators. Communities gain jobs but may lose the workers who kept other businesses running.

Property tax revenue increases dramatically. A seven billion dollar data center campus generates property tax payments measured in tens of millions annually. For a township with an existing budget of perhaps five million dollars, this represents transformational revenue. Schools can be built. Roads improved. Services expanded.

But the impact is also disruptive. Housing costs may rise as workers relocate or seek temporary accommodation. Construction brings heavy truck traffic on roads designed for farm equipment. Water and electricity consumption affects other users. A community's character changes when its largest employer becomes a featureless building full of computers.

Data centers create fewer permanent jobs per dollar of investment than almost any other industrial development. A one billion dollar manufacturing plant might employ a thousand workers. A one billion dollar data center might employ a hundred. The per-job cost of data center incentives often exceeds one million dollars—far higher than traditional manufacturing recruitment.[148]

Critics argue that communities surrender too much for too little employment. They point to tax exemptions that reduce revenue while providing minimal jobs. Construction jobs are temporary; infrastructure impacts are permanent. Defenders counter that data center jobs are well-paid and stable. Tax revenue matters independently of job counts—a school district benefits from property taxes whether the employer has one hundred workers or one thousand. Data centers provide economic diversification for communities previously dependent on agriculture or declining industries.

The debate cannot be resolved with simple metrics. Different communities have different priorities. Some value any economic development. Others prioritize employment over tax revenue. The right answer depends on local circumstances and community preferences.

Community opposition to data centers has grown significantly since 2020, drawing on multiple concerns: environmental advocates cite carbon emissions and water consumption; quality-of-life advocates cite noise and traffic; economic skeptics cite few jobs relative to investment.

Saline Township's consent agreement with Related Digital reflects this new reality. The agreement includes provisions for monitoring and mitigating impacts on neighboring wells and groundwater—acknowledgment that communities now demand specific protections rather than accepting promises.

7.9 THE FUTURE OF SITING

The farmland paradox will persist because its underlying economics are structural, not accidental. Greenfield sites are faster, cheaper, cleaner, and simpler. No realistic policy intervention will change these advantages. But three emerging trends may reshape siting patterns over the next decade.

Community resistance will become more significant. The 2025 wave of opposition across Michigan, Arizona, and other states signals that communities are no longer uniformly welcoming. Some projects will fail because no community will host them. Others will succeed by offering enhanced community benefits or choosing locations where opposition is weaker. The industry is adapting: more community engagement, more benefit agreements, more incorporation of local feedback into designs.

Water constraints will redirect investment. Water-stressed regions may become less viable as cooling demands increase. Water-rich regions—the Great Lakes states, the Pacific Northwest, parts of the Northeast—may capture a growing share. Air-cooled and closed-loop liquid cooling systems are improving, but evaporative cooling remains the lowest-cost option where water is available.

Brownfield development will remain niche under current incentives but could grow if policy changes. Sites with clean industrial histories—

printing, warehousing, light assembly—offer brownfield benefits without brownfield barriers. Sites with exceptional grid infrastructure, like former power plants, may justify remediation costs. These cases demonstrate that brownfield is possible under favorable conditions. Whether policy can create those conditions more broadly—through incentives that offset timeline penalties, through permitting reforms that level the playing field—is a question we return to in Chapter 13.

For Saline Township, the future holds transformation. *Harold*'s farm will become a data center campus. The township's tax base will expand dramatically. Some residents will benefit from employment and local business growth. Others will experience changed traffic patterns, altered community character, and the psychological weight of hosting a symbol of technological change they may not fully understand.

Harold drives past his former land occasionally, though his daughter wishes he would not. He takes the long way to the grocery store, the route that passes the property line where his grandfather planted the oak trees that are now scheduled for removal. He sees the survey stakes, the preliminary grading, the signs announcing what is to come. Security fencing has gone up where the farmhouse used to stand. They demolished it in November. He did not watch.

"It's strange," he says. He is sitting in the cab of his pickup truck, engine idling, heat running against the December cold. "Farming that land for fifty years. Watching my father farm it before me. Watching my grandfather plant those oaks because he wanted shade for generations that hadn't been born yet." He pauses. "And now it'll be something completely different. Something I don't even understand."

His daughter called from Chicago last week. She wanted to know how he was doing. He told her he was fine. He did not tell her about the drives, or the dreams where he is still farming, or the way he sometimes forgets and starts planning next year's rotation before remembering there will be no next year.

"But the world changes," he says. "The world always changes. My grandfather came here from Poland with nothing. Built something from nothing. Maybe that's what these people are doing too. Building something." He looks out the window at the construction equipment, the men in hard hats, the transformation underway.

"I just wish I knew what."

Ellen saw him there once, parked at the edge of what used to be his property, engine running against the February cold. His truck was the same Chevy he had driven for fifteen years, rust creeping up the wheel wells, the bed still stained with the memory of a thousand loads of feed and fertilizer. She pulled her own truck alongside his and rolled down her window. The cold air bit at her face.

"*Harold.*"

He looked at her without surprise. His face was thinner than she remembered, the lines deeper, the eyes more distant. They had known each other for forty years—she had taught his daughter biology at Saline High, watched his grandchildren at church picnics, shared equipment during harvest when one of them broke down. His hands on the steering wheel were the hands she had seen lift hay bales and fix machinery and hold his wife's hand at her funeral. Now they sat in their separate trucks, stopped in the road facing opposite ways, exhaust rising in twin plumes, looking at the construction site together.

"I should have held out," he said finally. "Like you."

"You couldn't know what was coming."

"I could have guessed." He rubbed his face with a thick hand. "Seven and a half million dollars. You know what I thought when I signed? I thought: my father would be proud. I thought: this is enough to take care of everyone." He laughed, but it was not amusement. "Now I drive past and I think: my father would have shot me."

"Your father would have understood," *Ellen* said, though she was not sure that was true. "Your kids weren't going to farm it. Neither were mine. The world changed."

"The world changed," *Harold* repeated. He watched a crane swing a massive steel beam into place, the metal catching the gray winter light. "But we're the ones who have to live in it."

They sat in silence for a while, engines idling, exhaust drifting across the frozen ground. The construction noise was constant—the beeping of trucks backing up, the clang of metal, the roar of earthmovers. Somewhere beneath that noise, if you listened hard enough, you could almost hear the land remembering what it used to be. But that was probably just imagination. The land did not remember. Only people did.

"Come over for dinner Sunday," *Ellen* said finally. Her voice came out rougher than she intended. "You shouldn't be alone so much. I'm making pot roast."

Harold nodded, though they both knew he probably would not come. He had not been the same since his wife died, and this—watching his life's work become something else entirely—had pushed him further into himself. His eyes were wet, but he did not wipe them. Neither of them mentioned it.

"I'll try," he said. It was the most he could promise. She reached over and touched his arm, just for a moment, her hand cold against his jacket. Then she rolled up the window and drove away, watching him in her rearview mirror until she turned the corner. He was still sitting there, engine running, looking at what used to be his farm.

The transformation of American farmland into data center campuses is one part of that change. Land use is shifting, driven by the economics of computation, the physics of electricity, and the choices of communities facing pressures they never anticipated. The paradox remains: brownfield sites wait empty while cornfields disappear under concrete. Policy has not resolved it. Markets will not resolve it either. The paradox will persist as long as greenfield advantages exceed the social and environmental costs of farmland conversion.

Whether that calculus ever changes depends on choices yet to be made—by policymakers, by communities, by the companies investing trillions of dollars in infrastructure that will shape America for generations to come.

The student in Ann Arbor finished her coffee an hour ago. The cup sits empty on the table, a ring of dried foam at the rim. She has moved on to other questions, other answers, other tokens flowing through infrastructure she has never considered. The conversation feels weightless, instantaneous, free.

Her question landed on *Harold's* farm.

Not literally—the signals traveled through fiber to servers in Virginia or Texas. But the demand her question represents, multiplied by hundreds of millions of users, created the market that brought Related Digital to Saline Township. The appetite for AI that makes ChatGPT possible is the same appetite that made *Harold's* land worth more as a data center than as a farm. His grandfather cleared those fields in 1922. His father expanded through the Depression. *Harold* himself survived the farm crisis of the 1980s. Now the land grows servers instead of soybeans.

We have traced her token through inference mathematics, through silicon fabrication, through the buildings that house the chips. We have followed the electricity from transformer to turbine. Now we see where that infrastructure touches ground—literally. The farmland paradox ensures that AI's physical footprint falls on agricultural communities, not abandoned factories. The land is bought. The fields are cleared. The transformation is underway.

But land alone does not build data centers. The Saline Township project costs seven billion dollars. The chips inside cost billions more. Construction requires thousands of workers. Operations require hundreds. Where does the money come from? Who decides to spend it? And what do they expect in return?

CHAPTER EIGHT

The Money

O N SEPTEMBER 4, 2024, Blackstone announced the acquisition of AirTrunk, an Australian data center operator, for sixteen billion dollars.[149,150] Stephen Schwarzman, the chairman and chief executive, had built Blackstone into the world's largest alternative asset manager, with over 1.2 trillion dollars in assets under management. This was the largest data center deal in history.

David watched the announcement from his office in Manhattan, thirty floors above Park Avenue. The September light slanted through floor-to-ceiling windows, catching the dust motes suspended in recycled air. His coffee had gone cold an hour ago. The Nespresso machine in the kitchenette down the hall hummed constantly—the only sound on a floor where thirty people worked in near-silence, staring at Bloomberg terminals and Excel models.

Fifteen years in private equity, eight at a data center-focused fund, and he had never seen anything like this. AirTrunk. Sixteen billion. A premium that made his colleagues gasp.

He pulled up his models, the familiar grid of cells that had guided a decade of decisions. His eyes were dry from too many hours of screen light. If Schwarzman was paying this much for Australian assets, what did that imply for American facilities? For the deals his fund was already circling?

The math said one thing. His gut—a tightness below his sternum that he had learned to trust—said another. This was either the beginning of the greatest infrastructure investment cycle in history or the top of a market that would make fools of everyone who chased it.

Blackstone was not simply buying a company. It was placing a bet on the future of global infrastructure. "Current expectations are that there will be approximately one trillion dollars of capital expenditures in the United States over the next five years to build and facilitate new data centers," Schwarzman explained at the firm's second-quarter earnings call, "with another one trillion dollars of capital expenditures outside the United States."[151] Blackstone intended to become the largest financial investor in AI infrastructure globally.

The AirTrunk acquisition followed Blackstone's ten billion dollar purchase of QTS Realty Trust three years earlier.[152] Combined with joint ventures with Digital Realty and other operators, the firm had assembled an eighty billion dollar data center portfolio with another hundred billion in development.[153] The thesis was simple: artificial intelligence would transform every industry. AI required computing infrastructure. Computing infrastructure required data centers. Data centers required capital at a scale only a handful of investors could provide.

"I believe the consequences of AI are as profound as what occurred in 1880 when Thomas Edison patented the electric light bulb," Schwarzman said. "While it took years to develop commercially viable products, the subsequent build out of the electric grid over the following decades has parallels to the creation of data centers today to power the AI revolution."[151] The technology was ready. The infrastructure was not. Someone would have to build it.

That someone, increasingly, was private equity.

8.1 THE TRILLION-DOLLAR BUILDOUT

The numbers are staggering. By the end of 2025, documented investment in United States data center infrastructure exceeded 1.1 trillion dollars across more than six hundred projects.[8] Average project size: 4.76 billion dollars. Median: 1.1 billion. The largest—Project Jupiter, the Stargate Santa Teresa development in New Mexico—represented 165 billion dollars of planned investment at a single location.

David keeps a spreadsheet on his laptop, a document that has grown to 847 rows across 23 tabs. He updates it weekly, usually on Sunday nights after his daughters are in bed, the house quiet except for the dishwasher cycling and his wife's footsteps upstairs. The numbers stopped making intuitive sense months ago. When he started in this business, a five hundred million dollar deal was significant. Now five hundred million is a rounding error.

"I tell my kids I work in finance," he says. His older daughter, eleven, asked him last month what he actually does. He tried to explain. She looked at him the way children look at adults who have stopped making sense. "They ask me what I finance. I say data centers. They ask how big. I say a trillion dollars." He shrugs. "They think I'm joking."

―――――

Five years earlier, these figures would have seemed impossible. In 2020, a billion-dollar data center was exceptional. By 2025, billion-dollar projects had become routine, ambitions measured in tens of billions. The transformation happened so fast that even industry veterans struggled to comprehend the scale.

To put these numbers in context, consider what else costs a trillion dollars. The entire United States interstate highway system, built over decades, cost roughly five hundred billion dollars in inflation-adjusted terms.[154] The Apollo program that put humans on the moon cost about

197

260 billion dollars adjusted.[155] The American Recovery and Reinvestment Act of 2009, the largest fiscal stimulus in history at that time, totaled eight hundred billion.[156]

The data center buildout is roughly equivalent to two interstate highway systems, constructed in under five years. And it is funded almost entirely by private capital, without the government debt that public infrastructure typically requires. That framing came under pressure in late 2025, when OpenAI's CFO floated the idea of federal loan guarantees to lower financing costs. The White House AI czar, David Sacks, responded bluntly: "no federal bailout for AI."[157]

Political declarations rarely survive contact with systemic risk. Consider the math. A trillion dollars of investment, financed at typical private equity leverage of sixty percent debt, means six hundred billion dollars in loans sitting on bank balance sheets and in pension portfolios. The Troubled Asset Relief Program that Congress passed in 2008—after initially rejecting it, then watching the stock market crash 777 points in a single day—authorized seven hundred billion dollars. The numbers are comparable.

The exposure runs deeper than debt. Microsoft, Apple, NVIDIA, Alphabet, Amazon, and Meta together represent roughly thirty percent of the S&P 500 by market capitalization. A typical 401(k) target-date fund holds substantial positions in all of them. When NVIDIA's stock price tripled in 2024 on AI demand, retirement accounts across America rose with it. The connection is not abstract: the same infrastructure buildout that transforms Saline Township also funds retirements in Saline Township. If AI investment collapses, the losses would propagate through every index fund in the country.

If AI becomes as embedded in the economy as its promoters claim, and if the investment thesis proves wrong, Congress may face the same choice it faced with the banks: let critical infrastructure fail, or bail it out. The precedent from 2008 suggests which way that choice goes.

The growth curve is stark. From 2010 through 2020, the industry documented $48.6 billion across 29 projects. Then the pace accelerated: $57.8 billion in 2021–2022, $29.0 billion in 2023. Then came the explosion: $276.3 billion in 2024 across 70 projects, $616.0 billion in 2025 across 56. Total documented investment now exceeds $1.1 trillion across 236 projects with disclosed figures.

Data center finance has evolved through four distinct phases. Each brought a different scale of capital, a different type of investor, and a different set of expectations about risk and return. Tracing these phases shows not just where the money comes from, but why the current investment surge may or may not be sustainable.

8.2 THE REIT ERA

David started his career in 2009, fresh from Wharton. His first job was at a real estate investment bank, analyzing shopping malls. The malls were dying. The financial crisis had gutted consumer spending. He needed a way out.

A colleague mentioned data centers. "Boring," he said. "Steady." That sounded perfect.

For most of the 2010s, data centers were a niche real estate play. A handful of publicly traded Real Estate Investment Trusts—companies that own income-producing properties and let investors buy shares rather than whole buildings—dominated the market: Equinix, Digital Realty, CyrusOne, QTS, CoreSite. They built facilities in major metropolitan areas, leased space to enterprise customers, and generated steady returns through long-term contracts with creditworthy tenants. The business was stable but unspectacular.

The economics were straightforward. A REIT would acquire land, build a facility, lease it to corporate customers at rates sufficient to cover construction costs and generate returns, then use the cash flow to finance the next building. Valuation multiples ranged from twelve to eighteen times EBITDA—earnings before interest, taxes, depreciation, and amortization, a measure of how much cash the business generates from operations—healthy for real estate but unremarkable compared to technology companies.

The REIT structure itself shaped the industry. REITs must distribute at least ninety percent of taxable income to shareholders, limiting retained earnings for reinvestment. They raise growth capital through equity offerings, which dilute existing shareholders, or debt, which increases financial risk. These constraints kept project sizes manageable and growth rates modest.

Typical deals ran from one hundred million to five hundred million dollars. Equinix might acquire a competitor's facility for three hundred million. Digital Realty might announce a five hundred million dollar development in Northern Virginia. AWS might reveal a billion-dollar regional buildout. These sums seemed substantial at the time. They would later seem quaint.

David remembers his first data center deal. A two hundred million dollar acquisition of a facility in New Jersey. He spent four months on due diligence, poring over every lease, every mechanical system, every utility contract. The closing dinner was at Smith & Wollensky on 49th Street, the red leather booths and wood-paneled walls that had witnessed a thousand such dinners. His managing director ordered the porterhouse for two. *David* was twenty-eight years old, still junior enough to feel the weight of the moment. "We opened champagne," he recalls. "For two hundred million. I remember thinking: this is it. This is what success feels like." He pauses, something flickering across his face. "Now we do two hundred million before breakfast. The champagne stopped meaning anything a long time ago."

––––––––

The REIT era established the foundational business model that still shapes the industry. Data centers are leased by the megawatt, with pricing typically ranging from one hundred to two hundred dollars per kilowatt per month for wholesale space. A ten-megawatt lease at one hundred fifty dollars per kilowatt per month generates eighteen million dollars in annual revenue. Contracts run five to fifteen years, providing predictable cash flows that support debt financing.

The REIT structure worked well for enterprise colocation—providing space, power, and cooling for corporate servers that companies didn't want to manage themselves. Fortune 500 companies outsourced their data centers to professional operators. Cloud computing pioneers like AWS leased space while building their own facilities. The market grew steadily at ten to fifteen percent annually.

But the model had limitations. REITs raised capital through equity and debt offerings, bounded by their existing portfolios and credit ratings. They built facilities speculatively or with anchor tenant commitments—major customers who sign long-term leases, reducing financing risk—limiting individual project sizes. They operated in a market where demand grew steadily but not explosively, keeping valuations modest. This conservative approach served the industry well through the 2010s, producing steady returns and stable asset values. It could not scale to meet the explosive demand that AI would generate.

By 2020, the total documented investment in data centers from the REIT era amounted to roughly 48.6 billion dollars across twenty-nine disclosed projects. It was a respectable sum. It would be dwarfed by what came next.

8.3 PRIVATE EQUITY ARRIVES

The shift began in 2021, when Blackstone acquired QTS Realty Trust for ten billion dollars.[152] The deal represented a twenty-one percent premium to QTS's trading price and valued the company at approximately twenty-five times EBITDA—a significant expansion from the twelve-to-eighteen-times multiples that had characterized the REIT era.

David's phone rang at 6 AM the morning the deal was announced. He was standing in his kitchen in Westchester, still in the t-shirt he had slept in, waiting for the coffee maker to finish its cycle. Gray light through the window. His wife and daughters still asleep upstairs. The caller ID showed his managing partner, calling from JFK.

"Did you see QTS?"

"Twenty-five times," *David* said. His voice came out rough, not yet warmed up for the day. "Blackstone's lost their minds."

"Or they know something we don't." *

David spent that week rebuilding every model. He worked until midnight most nights, the office emptying around him, the cleaning crew vacuuming between the cubicles while he stared at sensitivity tables. If twenty-five times was the new floor, what was the ceiling? His fund had been acquiring data center assets at fourteen to sixteen times. Suddenly their portfolio looked undervalued. Or the market had gone crazy. Possibly both.

The premium signaled a fundamental revaluation. Blackstone was paying for optionality: the ability to accelerate growth beyond what public markets would tolerate, to capture value from trends that quarterly-focused investors were underappreciating. Private ownership would allow patient capital, aggressive development timelines, and operational improvements that public company reporting cycles made difficult. The firm brought not just money but expertise in infrastructure development, relationships with utilities and construction firms, and willingness to build speculatively rather than wait for anchor tenants.

The thesis proved prescient. Under Blackstone ownership, QTS expanded its leased capacity nine-fold—from conservative development to aggressive expansion, building facilities in anticipation of demand rather than waiting for signed leases. That growth would have been impossible under public company constraints.

The QTS deal triggered a wave of take-private transactions. KKR and Global Infrastructure Partners acquired CyrusOne for fifteen billion in March 2022.[158] DigitalBridge and IFM Investors took Switch for eleven billion later that year.[159] Brookfield acquired Compass Datacenters and other platforms.[160] Suddenly, the largest names in private equity were competing for data center assets. Take-private value exceeded forty billion dollars in just two years.

Why did private equity firms pay such large premiums to acquire public REITs? Three factors drove the math.

First, low interest rates made debt cheap. A deal financed at four percent interest has very different economics than the same deal at eight percent. The 2021–2022 environment offered historically favorable financing.

Second, cloud computing demand was accelerating faster than public markets projected. The hyperscalers—AWS, Microsoft Azure, Google Cloud—were expanding capacity at remarkable rates. Every major corporation was moving workloads to the cloud. The demand trajectory justified aggressive bets.

Third, private ownership enabled operational changes—accelerated development, reduced reporting burden, strategic repositioning—that public markets resisted. Public company executives face quarterly earnings pressure, analyst scrutiny, shareholder activism. Private executives can make long-term investments without explaining every decision to outside shareholders.

The sponsors believed they could create value faster than public shareholders demanded, then sell or re-IPO at even higher multiples.

Valuations expanded accordingly. CyrusOne's fifteen billion dollar price implied roughly twenty-one times EBITDA. Switch pushed higher still. Investors who had paid twelve to fifteen times EBITDA for data center assets in the mid-2010s now saw multiples approaching twenty-five times for premium platforms.

Total investment during 2021–2022 reached 57.8 billion dollars across twenty-four projects—more than the entire preceding decade in just two years. But even this acceleration was merely prologue.

8.4 THE AI TRANSITION

The year 2023 was transitional. ChatGPT had launched in November 2022, demonstrating that large language models could generate useful, coherent text in response to natural language prompts. Within months, every major technology company was racing to build or acquire AI capabilities. AI required computing infrastructure at massive scale.

David used ChatGPT for the first time in January 2023. His daughter—the older one, then nine—had shown him over Christmas break, sitting on the couch in the living room while snow fell outside and his wife read in the armchair by the fire. "Dad, look at this." She typed a question about dolphins. The response appeared, fluent and confident. *David* took the laptop. He asked it to explain leveraged buyouts. The response was decent—not perfect, but decent. He asked it to draft a term sheet. Better than some junior analysts.

He sat there for an hour after his daughter lost interest and wandered off, the fire dying to embers, his wife asking if he was coming to bed. He could not stop thinking about what it would take to run this at scale. The servers. The power. The cooling. The real estate.

"This changes everything," he told his partners the following week, standing at the head of the conference room table, the skyline of midtown Manhattan visible through the windows behind him. "I don't know if AI will take over the world. But I know it needs data centers. A lot of them."

———

GPU shortage hit first, with lead times stretching to eighteen months for large orders. Companies that had ordered early suddenly possessed strategic assets worth thirty to forty thousand dollars per unit—if you could get them. Companies that had not scrambled on secondary markets or rented capacity from the few providers that had it. Data center operators with AI-ready facilities discovered their assets were worth far more than traditional colocation space. High-density power, liquid cooling, low-latency networking: these capabilities now commanded premium prices.

The shortage revealed a bottleneck. Training large language models requires thousands of GPUs working together, each consuming significant power and generating significant heat. Traditional data centers, designed for enterprise servers at five to fifteen kilowatts per rack, cannot accommodate AI workloads at forty to one hundred kilowatts per rack. The facilities simply did not exist at the scale required.

CoreWeave exemplified the transition. The company had started as a cryptocurrency mining operation, accumulating NVIDIA GPUs to mine Ethereum. When cryptocurrency prices collapsed, CoreWeave pivoted to cloud computing, offering GPU access to AI researchers and companies that could not secure their own hardware.

The pivot proved extraordinarily well-timed. CoreWeave had GPUs. It had high-density power infrastructure. It had cooling systems. When AI demand exploded, CoreWeave was one of the few operators ready to serve it.

NVIDIA invested directly, taking an equity stake that validated CoreWeave's model. The move was strategic: NVIDIA needed customers who could actually deploy its chips at scale. Most enterprises could not. CoreWeave could. Its valuation rose from negligible to twenty billion dollars in under two years.

The investment numbers reflected this transition. Total deployment in 2023 reached 29.0 billion dollars across eighteen projects—down from 2022 but representing a shift toward AI-specific infrastructure. Average project size climbed as developers built for the new workload requirements. The first truly massive AI infrastructure projects began to take shape.

Brookfield's acquisition of Compass Datacenters for 5.5 billion dollars in December 2023 signaled that infrastructure funds remained committed to the sector. But the character of the investment was changing. Facilities designed for general-purpose computing were giving way to specialized AI training and inference infrastructure. The next phase would bring investment at a scale that made even the 2021–2022 take-private boom look modest.

8.5   THE MEGA-DEAL ERA

By 2024, data center investment had entered a new regime. Individual projects routinely exceeded ten billion dollars. Consortium structures brought multiple capital sources together for deals too large for any single sponsor. Hyperscalers—the technology giants that actually used the computing capacity—began investing directly at scales that dwarfed traditional financial sponsors.

David's fund had a two billion dollar mandate. He used to feel that was substantial. Now he sat in meetings where two billion was the equity check for one project, with five other funds writing similar checks. "The room is different now," he says. "Used to be we'd compete against two or three bidders. Now there are twelve. Sovereign wealth funds. Infrastructure funds. Hyperscalers. Everyone."

The transformation was quantitative and qualitative. More money flowed in, but the projects themselves had changed. Where 2020-era facilities housed enterprise servers for dozens of corporate customers, 2024-era facilities might house AI training clusters for a single hyperscaler. Economics, engineering, business models—all had evolved.

The numbers tell the story. In 2024, seventy disclosed projects attracted 276.3 billion dollars—more than four times the 2021–2022 total in a single year. Average project size: 3.9 billion dollars. Eight projects exceeded ten billion. Three exceeded twenty billion. The industry had transformed.

The next year accelerated further. Through the first ten months, fifty-six projects attracted 616.0 billion dollars. Average project size reached eleven billion. Twenty-four projects exceeded ten billion. Nine exceeded twenty billion. Three exceeded fifty billion. The market's largest deals would have seemed incomprehensible just four years earlier. Now they represented the logical extension of a market growing at exponential rates.

The concentration of capital is striking. Just eleven projects with investment exceeding twenty billion dollars represent 45.2 percent of total documented investment. The top twenty projects account for 59.5 percent. Data center finance has become extraordinarily concentrated, with a handful of mega-deals dominating the market.

The deals required new financial structures. No single private equity firm, however large, could fund a twenty billion dollar project alone. Consortiums emerged: multiple equity sponsors, hyperscaler anchor tenants providing committed revenue, utility partners guaranteeing power delivery, debt providers across multiple tranches, sometimes sovereign wealth funds seeking stable infrastructure returns.

David describes a recent negotiation. Seven parties. Three time zones. Fourteen lawyers in one room, plus video screens showing another dozen. "We argued for six hours about a single clause," he says. "The power commitment guarantee. If the utility fails to deliver, who bears the risk? The operator? The anchor tenant? The equity sponsors pro rata? The lenders?"

He laughs, though it does not sound like amusement. "We settled at 2 AM. I'm not even sure what we agreed to. The lawyers are still drafting."

————

The Stargate initiative illustrated the consortium model at its most ambitious. OpenAI, Oracle, SoftBank, and NVIDIA together committed to a five hundred billion dollar AI infrastructure program, with an initial forty billion dollar facility in Abilene, Texas.[2,55] No single company could have financed it alone. Together, they assembled capital at a scale that redefined what was possible.

The Stargate announcement reflected several trends converging. OpenAI needed computing capacity to train and deploy ever-larger models. Oracle needed a market for its cloud infrastructure ambitions. SoftBank's Masayoshi Son had long bet on transformative technology. NVIDIA needed customers at scale for its chips. Each partner brought something the others needed.

Whether Stargate would achieve its stated ambitions remained uncertain. Five hundred billion dollars across multiple facilities over multiple years required sustained execution at extraordinary scale. Power infrastructure had to be built. Communities had to be convinced. Construction had to proceed on schedule. Dozens of variables could derail the plan.

But the announcement itself shaped market expectations. If Stargate was the benchmark for AI infrastructure ambition, other players would need to respond with comparable scale. The competitive dynamic pushed investment higher.

8.6 WHO PROVIDES THE CAPITAL

The current investment environment draws on several distinct capital sources, each with different expectations, time horizons, and strategic motivations. The mix has shifted dramatically as project sizes have grown.

The technology giants—Microsoft, Amazon through AWS, Google through Alphabet, and Meta—collectively spend more on data center infrastructure than any other category of investor. Microsoft's capital expenditure budget reached eighty billion dollars annually by 2024, the majority dedicated to AI infrastructure supporting its cloud business and OpenAI partnership.[9] AWS committed over seventy-five billion per year.[161] Google announced fifty billion.[162] Meta budgeted thirty-seven to forty billion annually.[163]

Combined hyperscaler spending exceeds two hundred billion dollars annually—dwarfing private equity and infrastructure fund investments. Hyperscalers build for their own use, with specifications tailored to their workloads; financial returns matter less than strategic positioning. Their buildout also creates opportunities for other investors: they lease significant capacity from operators like Equinix and Digital Realty, and their anchor tenant commitments enable private equity sponsors to finance new developments.

209

Private equity firms approach data centers as financial engineering opportunities: acquire platforms, improve operations, accelerate growth, exit at higher valuations. Target returns run twenty to twenty-five percent IRR—internal rate of return, a measure of expected annual investment yield—over four to seven year holds, achieved through operational improvements, growth acceleration, multiple expansion, and debt reduction.

David thought he understood IRR after his first week at Wharton, but he never really felt it in his bones until his third year at Goldman. A professor can explain that IRR is the discount rate making net present value equal zero. A managing director, chain-smoking on a rooftop at two in the morning, explains what that actually means: "If I give you a dollar today, and you give me back a dollar twenty-five in one year, I made twenty-five percent. Simple. But what if you give me back nothing for four years, then sixty dollars all at once? What did I actually earn per year?" IRR answers that question—it translates lump-sum returns into annual equivalents, so investors can compare a four-year bet against a seven-year bet against a ten-year hold. The math gets complicated. The intuition stays simple.

David explains the logic. "We're not building infrastructure for the public good. We're building it to make money for our investors. Pension funds. Endowments. Insurance companies. They give us capital; we owe them returns."

He pauses. "Does that sound cold? It's just reality. The money has to come from somewhere. The money has to go somewhere."

Blackstone dominates with its 80 billion dollar portfolio and 100+ billion dollar development pipeline, combining platform acquisitions like QTS and AirTrunk, joint ventures with Digital Realty and COPT, power infrastructure investments in Invenergy and PPL, and technology bets like the CoreWeave debt facility. KKR and Global Infrastructure Partners brought fifteen billion to CyrusOne. DigitalBridge has deployed roughly twelve billion across Vantage, DataBank, and Switch. Together, private equity firms control portfolios valued well over one hundred billion dollars.

NVIDIA has emerged as a new category of investor entirely, with over one hundred billion dollars committed—including a hundred billion to OpenAI alone, plus five billion to Intel and billions more in equity stakes across CoreWeave, Applied Digital, and other infrastructure providers.[164] By providing equity, debt, and vendor financing to operators who buy its GPUs, NVIDIA has created a novel form of vertical integration—securing distribution channels while accelerating deployments that might otherwise stall on capital constraints.

Infrastructure funds seek twelve to fifteen percent returns over ten to fifteen year holds, treating data centers as essential utilities analogous to toll roads or airports. Consider Macquarie Asset Management's 2025 sale of Aligned Data Centers at a forty billion dollar enterprise value.[165] Macquarie had transformed Aligned from a two-facility operator into a fifty-campus Americas platform. That sale demonstrated the potential for patient infrastructure capital. GIP, now part of BlackRock following its 2024 acquisition, participated in the CyrusOne transaction.[166] Brookfield Infrastructure and Stonepeak have built substantial portfolios. These investors emphasize operational improvement over financial engineering, holding assets for decades when cash flows justify patience.

Publicly traded REITs remain significant owners, though their relative importance has declined as private capital scaled up. Equinix, with over 260 data centers globally, operates primarily as a colocation provider focused on network interconnection.[167] Digital Realty has evolved to embrace joint venture structures, partnering with Blackstone and others to fund developments that would strain its balance sheet alone. Smaller public data center REITs have largely been acquired by private equity or strategic buyers.

————

Sovereign wealth funds have emerged as significant participants, particularly for the largest projects. The most notable entry came in 2024, when MGX—the UAE's technology-focused investment vehicle—joined BlackRock, Global Infrastructure Partners, Microsoft, and NVIDIA in the AI Infrastructure Partnership.[168] The consortium targets one hundred billion dollars in data center investment, beginning with its forty billion dollar acquisition of Aligned Data Centers.[165] Singapore's Temasek and GIC, Saudi Arabia's Public Investment Fund, and Canada's pension funds have also taken positions.

Sovereign wealth capital brings unique characteristics: indefinite time horizons, modest return requirements where eight to ten percent may satisfy mandates, and capital pools enormous enough to write checks no private equity firm could deploy from a single fund. For mega-projects requiring tens of billions, sovereign wealth participation can be decisive.

8.7 DEAL STRUCTURES

The evolution from hundred-million-dollar facilities to fifty-billion-dollar mega-projects required parallel evolution in deal structures.

In the traditional model of 2010–2020, a single buyer acquired or developed a facility with forty to fifty percent equity and the balance in senior secured debt. A five hundred million dollar deal could be financed by a single fund. A fifty billion dollar deal required an entirely new approach.

Take-private transactions in 2021–2022 brought more complex financing. The QTS acquisition combined Blackstone equity with significant leverage; leveraged buyout financing from term loan B providers and high-yield bond markets enabled deal sizes that traditional financing could not support. These structures worked for established platforms with predictable cash flows but could not easily finance development projects for facilities that did not yet exist.

––––––

Mega-projects require consortium structures because no single capital source can prudently commit tens of billions to a single development. Today's largest deals combine multiple elements.

Anchor tenant commitments from hyperscalers provide revenue certainty that supports debt financing. *David* puts it bluntly: "Without an anchor, you can't get a loan. Period." An anchor tenant—typically a hyperscaler like Microsoft or AWS—signs a long-term lease before construction begins, guaranteeing revenue for ten or fifteen years. That commitment transforms a speculative real estate project into something lenders can underwrite. "The bank doesn't care about your growth projections," *David* says. "They care about counterparty risk. Who's on the other side of the contract? Microsoft? Great, they're investment-grade. Some startup nobody's heard of? Good luck."

Multiple equity sponsors share the equity commitment, typically two to five funds per deal. Vendor financing from NVIDIA and others reduces capital requirements while securing demand. Utility partnerships guarantee power delivery in exchange for long-term commitments. Layered debt spans senior secured, high-yield, and mezzanine tranches. Sovereign wealth provides patient capital with indefinite time horizons.

The result: capital stacks with five to ten distinct sources, each with different expectations, governance rights, and exit timelines. A single consortium financing might involve twenty separate agreements among a dozen parties. Legal fees alone can exceed fifty million dollars.

These structures enable projects at scales otherwise impossible. The largest developments exceed the entire 2010–2020 investment total, requiring consortium financing on an exceptional scale.

8.8 VALUATION

Data center valuations have expanded from twelve times EBITDA to fifty times or more—one of the most dramatic repricing events in infrastructure history. Understanding why requires examining what buyers actually purchase.

David thinks about valuation constantly. His fund's entire business depends on getting it right.

"In the REIT days, valuation was simple," he says. "Cash flows. Growth rates. Risk premiums. You could build a model, defend it, argue about the assumptions." He taps the table. "Now? Now we're pricing things that don't exist yet. Capacity that hasn't been built. Demand that might not materialize. It's educated guessing dressed up in spreadsheets."

In the REIT era, data centers traded as real estate with long-term tenants. Valuation reflected the net present value of contracted cash flows, plus reasonable assumptions about renewal rates and rent escalations. The math resembled industrial or office buildings with creditworthy tenants.

The valuation methodology began with existing contracts. A facility with ten megawatts leased at one hundred fifty dollars per kilowatt per month generates eighteen million in annual revenue. At seventy percent EBITDA margin, that yields 12.6 million in EBITDA. At fifteen times EBITDA, the facility is worth 189 million dollars.

Twelve to eighteen times EBITDA implied discount rates of eight to twelve percent and modest growth assumptions. Buyers earned returns through yield—the ongoing cash distributions from existing contracts—

plus modest appreciation from portfolio growth and rent increases. This valuation approach rewarded stable operations and predictable cash flows.

Private equity take-privates paid premiums reflecting several factors beyond existing cash flows:

Private ownership could accelerate development faster than public markets tolerated, creating value public markets had not priced. Without quarterly earnings pressure, management could focus on long-term value rather than short-term metrics, potentially improving margins and returns.

Low interest rates reduced the cost of leverage, mathematically supporting higher entry prices. The same cash flows could support more debt, allowing sponsors to pay higher prices while maintaining equity returns. And limited acquisition targets in a sector with strong fundamentals commanded competitive pricing. When multiple buyers pursue limited targets, prices rise.

Twenty to twenty-five times EBITDA remained connected to existing cash flows but incorporated more aggressive growth assumptions. Returns would come partly from yield, partly from appreciation as portfolios expanded.

The AI era has changed valuation logic fundamentally. Buyers now pay for capacity rather than cash flows—the ability to deploy GPUs and run AI workloads matters more than current tenant contracts. Facilities with AI-ready infrastructure—high-density power, liquid cooling, low-latency networking—command dramatic premiums over traditional colocation space.

The shift reflects scarcity. AI workloads require specialized infrastructure that most facilities cannot provide. Traditional data centers cannot simply be upgraded; the power densities, cooling requirements, and networking specifications differ fundamentally. Building new AI-ready facilities takes years. Demand is immediate. The imbalance drives prices.

CoreWeave's valuation trajectory illustrates the shift. The company raised equity at nineteen billion dollars in early 2024, reached twenty-three billion by late 2024, and went public at roughly twenty-seven billion in March 2025.[169,170] This valuation exceeded any traditional multiple analysis. It reflected scarcity: AI computing capacity in a market where every major technology company needed more than it could secure.

Fifty times EBITDA—or higher for pre-revenue facilities—means buyers purchase strategic options, not income streams. The bet: AI demand will grow fast enough and long enough to justify prices that make no sense under traditional real estate math.

The logic has internal consistency. If AI computing demand doubles annually, today's expensive capacity becomes cheap in two years. If the alternative is having no capacity at all, paying high prices beats losing market position. If competitors pay similar prices, declining to participate means falling behind.

But the logic depends on assumptions that cannot be verified. Demand must continue growing. Supply must remain constrained. The technology must not shift to architectures that make current infrastructure obsolete. Many things must go right for current valuations to prove justified.

Are these valuations justified or bubble pricing? The honest answer: no one knows. If AI computing demand continues doubling annually for several more years, current prices may prove cheap in retrospect. The facilities built today will be essential infrastructure for decades. If demand plateaus or efficiency improvements reduce computing requirements, current prices will prove ruinous—investors will have paid fifty times EBITDA for assets worth fifteen times. Some will be right. Others will lose billions. The outcome depends on technological and economic developments no one can reliably predict. Concentration of capital creates concentration of risk. The trillion-dollar buildout represents a trillion-dollar bet on AI demand. If the bet is wrong, the losses spread across the investors who funded it.

8.9 RELATED DIGITAL

Who is building the Saline Township facility, and why?

Related Companies, founded by Stephen Ross in 1972, is one of America's largest real estate developers. The company built Hudson Yards in Manhattan, the most expensive real estate development in American history. It has developed over sixty billion dollars of real estate across residential, commercial, and mixed-use projects.

Ross exemplifies the American real estate developer archetype. Born in Detroit, educated at the University of Michigan and New York University, he built Related from a single affordable housing project into a diversified development empire. Forbes estimates his net worth at over eight billion dollars.

Ross's Michigan connections run deep. He has donated hundreds of millions to his alma mater, funding the business school and athletic facilities. When Related Digital sought sites for AI infrastructure, Michigan was a natural choice.

Related Digital, the company's data center division, represents a bet that Related's real estate expertise can extend to computing infrastructure. The skills that built Hudson Yards—land assembly, construction management, tenant relations, capital formation—may translate to data center development. Or they may not. Related is learning as it goes.

The Saline project, announced as a partnership with Oracle and OpenAI as part of the broader Stargate initiative, represents Related's largest
data center bet to date.[18]

The deal structure reflects consortium financing principles. Oracle
serves as anchor tenant, providing committed demand that supports development financing. Oracle's credit rating allows Related to secure debt
at favorable rates. OpenAI's involvement signals the facility will support
frontier AI training and inference workloads—the highest-value use case
for AI infrastructure.

Related brings development expertise and capital. The company
knows how to assemble land, manage construction, and deliver complex projects. These skills transfer from office buildings to data centers,
though the technical requirements differ significantly.

DTE Energy's power commitments, described in Chapter 4, are essential.[142] Without power, the data center cannot operate.

The University of Michigan connection runs deeper than Ross's
alumni status. The institution has separately announced a 1.2 billion dollar data center project with Los Alamos National Laboratory in neighboring Ypsilanti Township. The proximity is no coincidence. Michigan is positioning itself as an AI infrastructure hub, with multiple projects drawing
on state incentives, utility cooperation, and institutional relationships.

For Saline Township, the project represents Related Digital's bet that
southeastern Michigan can host gigawatt-scale AI infrastructure. For Related, it represents diversification into a sector with dramatically different
economics than traditional real estate. For the financial sponsors backing
the project, it represents an opportunity to participate in the AI infrastructure boom.

David's fund looked at Related Digital. They passed.

"Too early," he explains. "Related knows real estate. They don't know data centers. Not yet." He pauses. "But the deal structure was interesting. Oracle anchor. Power commitments locked. If they execute, it'll work."

He had followed the Saline Township story in the trade press—the initial rejection, the lawsuit, the eventual settlement. "That's the part people don't see," he says. "The community stuff. The politics. Everyone in this business focuses on the cap stack and the power agreements. But there's a township supervisor somewhere making decisions about something he's never seen before. Board members who probably don't understand what a gigawatt is." He shakes his head. "I wonder sometimes whether they know what they're approving. Whether anyone explained it to them in terms they could understand."

He pulls up a map on his laptop, zooms to southeastern Michigan. "Four hundred fifty permanent jobs. A township with what, two thousand people? Their budget is probably five million a year." He traces the project boundary with his finger. "The farmer who sold that land—he probably got more money than his family made in a century. Is that good? Is that fair? I don't know. It's the market. The market doesn't care about fair."

He closes the laptop. "That's the question with all of these. If they execute."

———

The economics matter for the township. Related Digital will pay property taxes on a multi-billion dollar facility. Those payments will transform township finances, potentially enabling investments in schools, roads, and services otherwise impossible. Construction will bring thousands of jobs. Operations will bring hundreds more.

But the community will also bear impacts—traffic, water consumption, changed character—that accompany hosting infrastructure at this scale. The consent agreement negotiated after Related's lawsuit includes provisions intended to mitigate these effects. Whether the mitigations prove sufficient remains to be seen.

8.10 RISK: WHAT COULD GO WRONG

The extraordinary valuations and capital commitments create extraordinary risks. Investors are betting that AI demand will justify prices that make no sense under any other scenario. Several categories of risk deserve attention.

David does not sleep well anymore.

"I used to sleep fine," he says. "The deals were smaller. The bets were manageable. Now?" He shakes his head. "Now I lie awake thinking about all the ways this could go wrong."

The entire investment thesis assumes AI computing demand will keep growing at exceptional rates for years. But demand projections are uncertain.

AI model efficiency improvements could reduce computing requirements per unit of output. Researchers find ways to achieve comparable results with smaller models. Inference optimization reduces the computing required to run trained models. If efficiency improves faster than capability demands increase, computing requirements could plateau.

Competitive pressure could compress the economics of AI services. As more providers enter the market, pricing may fall. Lower prices reduce data center revenue. Lower revenue reduces investor returns.

Economic slowdowns could reduce enterprise AI adoption. Companies invest in AI when they expect returns. Recession fears could delay those investments, reducing demand for computing capacity.

If demand grows more slowly than expected, hundreds of billions of dollars of infrastructure could become overcapacity. Facilities built for

2028 demand might sit partially empty if that demand arrives in 2032. Or never arrives at all.

_____

Current data centers are optimized for NVIDIA GPU-based computing. But computing architectures change. The history of computing is a history of architectural transitions: mainframes to minicomputers, minicomputers to personal computers, CPUs to GPUs. Each transition rendered previous infrastructure less valuable.

Neuromorphic chips, quantum computing, or breakthrough efficiency improvements could make current infrastructure less valuable or even obsolete. Facilities designed for 2024 workloads may not suit 2030 requirements.

The risk is particularly acute for specialized AI infrastructure. General-purpose data centers can serve multiple workload types. AI-specific facilities optimized for GPU computing may have limited alternative uses if GPU computing becomes obsolete.

David thinks about this often. "We're building for a technology that's six years old," he says, meaning GPU-accelerated neural networks. "What if something better comes along? What if these facilities are the data centers equivalent of mainframe rooms—impressive in their time, obsolete in a decade?"

He does not have an answer. No one does.

Policy risk compounds technology risk. *David* watched the export control regime shift three times in twelve months: the AI Diffusion Rule in January, its rescission in May, the H200 revenue-sharing approval in December. Each shift revalued assets differently.

"In January, we thought maximum restriction was the future," he explains. "Build domestically, serve domestic customers, let China figure it out. By December, the calculus changed. If H200s flow to China with a government cut, Chinese data centers compete with ours for some workloads. The revenue share helps NVIDIA's margins, but it changes the competitive dynamics for everyone else."

He pulls up a sensitivity table on his laptop, the familiar grid of cells that has guided so many decisions. "We model technology risk. We model demand risk. How do you model policy risk when the policy can flip completely between administrations?" The cell showing export control assumptions is highlighted in yellow—flagged for uncertainty that no historical data can resolve.

———

Many announced projects depend on power infrastructure that does not yet exist. Utility construction delays, regulatory approval challenges, or grid constraints could prevent projects from reaching their planned capacity.

Power infrastructure requires years to build. Transmission lines need rights-of-way that may face opposition. Substations need transformers that face supply chain constraints. Generation facilities need permits that may be delayed or denied.

Projects that cannot secure power cannot generate returns. A data center without electricity is an expensive building. Investors who committed capital based on power projections may find those projections unfulfilled.

———

Fifty times EBITDA multiples leave no room for disappointment. The valuation assumes everything goes right: demand grows, supply stays constrained, technology holds steady, execution proceeds smoothly.

If growth slows, multiples will compress—potentially dramatically. Investors who paid premium prices for AI optionality may find their holdings worth far less than anticipated.

The compression could be rapid. Public markets reprice assets in hours. Private valuations take longer to adjust but eventually reflect reality. An asset purchased at fifty times EBITDA might be worth fifteen times a year later if sentiment shifts.

Highly leveraged capital structures are vulnerable to interest rate increases. The 2024–2025 investment boom occurred as rates declined from 2023 peaks. If rates rise substantially, refinancing costs could squeeze returns on existing investments and make new deals less attractive.

Many data center acquisitions were financed with floating-rate debt. Interest payments rise with rates. Higher payments reduce cash available for other purposes, impairing the ability to invest in growth or service existing obligations.

Rising rates also affect new deal economics. Higher financing costs require either lower purchase prices or lower equity returns. Sellers resist lower prices. Sponsors resist lower returns. The result: fewer deals.

The risks examined so far fall primarily on investors. But data center development creates financial exposure for parties who did not choose to participate: utilities that build infrastructure, municipalities that provide services, ratepayers who fund the grid.

Utilities face significant exposure. When a utility commits to serve a gigawatt data center, it must invest in transmission lines, substations, and potentially new generation—investments reaching hundreds of millions of dollars that take years to complete. If the data center project scales back or cancels, the utility holds infrastructure designed for demand that never materialized.[171]

AEP Ohio cut its data center demand pipeline from over 30 gigawatts to 13 gigawatts after culling "opportunistic would-be developers who lack financial strength." The culled projects represented infrastructure commitments that could have become stranded assets. AEP subsidiaries in Indiana, West Virginia, and Kentucky had already acquired 750 megawatts of generating capacity for data centers that never materialized—capacity they now seek to sell into PJM's capacity auction.[172] The experience illustrates how speculative demand can evaporate, leaving infrastructure investments without the customers they were designed to serve.

Utilities are responding with increasingly protective contract structures. Virginia's State Corporation Commission approved Dominion Energy's new GS-5 rate class requiring fourteen-year minimum contracts, 85 percent minimum payment for transmission and distribution capacity, and exit fees for early departure.[173] Indiana Michigan Power's modified tariff requires twelve-year terms and 80 percent minimum monthly demand charges. These structures shift risk from utilities to data center operators—but only if operators honor their commitments.[174]

Ratepayers bear a growing burden. Grid upgrades built for data centers are paid through rate bases that include all customers. PJM's independent market monitor found data centers responsible for 63 percent of the price increase in the 2025/2026 capacity auction, translating to 9.3 billion dollars in additional costs passed to ratepayers across the thirteen-state region.[96] Washington D.C. Pepco customers saw bills rise twenty-one dollars monthly starting June 2025, with roughly ten dollars attributed to capacity market price spikes driven by data center demand.[175]

The Natural Resources Defense Council projects household costs could rise seventy dollars monthly through 2033 if current trends continue—163 billion dollars in additional ratepayer costs.[176] These costs materialize regardless of whether individual data center projects succeed. Ratepayers fund the infrastructure. Data center operators capture the returns.

Local governments face their own exposure through tax increment financing, infrastructure bonds, and services sized for projected growth. The Government Finance Officers Association warns that TIF creates "significant risk" when incremental assessed value fails to materialize.[177] If projected property values do not rise, municipalities may be unable to repay TIF bonds, shifting the burden to all municipal property taxpayers.

Thirty-seven states have passed data center-specific sales tax exemptions, with sixteen states granting nearly six billion dollars in exemptions over five years.[178] These exemptions represent foregone revenue that cannot fund other priorities. If data centers fail to generate the economic activity that justified the exemptions, the foregone revenue becomes permanent loss.

The telecommunications bubble of the late 1990s offers a cautionary template. Telecom companies invested more than 500 billion dollars in fiber optic infrastructure, mostly debt-financed. By 2002, only 2.7 percent of installed fiber was in use.[179] Twenty-three companies went bankrupt. Five hundred thousand workers lost jobs. Communities that had built around telecom hubs found themselves stranded.[180] The data center buildout differs in important ways: anchor tenant commitments, physical power constraints, measurable current demand.[181] But the lesson remains. When private investors bet aggressively on transformative technology, the costs of failure are not always borne by those who made the bets.

8.11 THE GEOGRAPHY

The geographic distribution of data center investment has shifted as project scale has grown. Traditional hubs—Northern Virginia, the Bay Area, Chicago, Dallas—are giving way to new locations offering gigawatt-scale power and abundant land.

The leading states by announced investment reveal the new geography: New Mexico leads with $167.2 billion, driven by Project Jupiter;

Kansas follows at $128.8 billion with Project Kestrel; Pennsylvania reaches $125.1 billion across multiple projects; Georgia $79.8 billion; Texas $78.2 billion; Arizona $63.4 billion; Virginia $56.6 billion.[8] The top three states capture 37.5 percent of total documented investment.

The shift reflects themes explored throughout this book. Power availability determines where mega-projects can locate. New Mexico's Project Jupiter draws on power resources unavailable in coastal metro areas. Kansas offers abundant land and utility cooperation. Pennsylvania combines Marcellus Shale natural gas with industrial-era transmission infrastructure.

Northern Virginia's Data Center Alley remains the largest existing concentration of data centers in the world, but the region is increasingly constrained—power availability, land scarcity, community opposition. Texas offers deregulated electricity markets and political support, though grid reliability concerns persist. The Mountain West has attracted growing investment for its cool climate and inexpensive land.

Michigan's emergence as a data center destination reflects these dynamics. The state offers power availability through DTE Energy, land in southeastern Michigan townships, water resources from the Great Lakes, and state incentives designed to attract investment. Saline Township and multiple other developments position Michigan to capture a meaningful share of the buildout.

8.12 THE FUTURE

Several trends will shape data center finance over the coming years.

Consolidation will continue. Smaller operators will be acquired by larger platforms seeking scale advantages in capital access, operational efficiency, and customer relationships.

Consortium structures will become standard for mega-projects. Fifty billion dollar developments require multiple capital sources. The expertise to structure such deals will become a differentiating capability.

Power infrastructure will increasingly bundle with data center investment. Blackstone's Pennsylvania strategy—combining data center development with natural gas generation through the PPL joint venture—previews vertical integration that creates competitive moats.

Returns will compress if supply catches up to demand. Current prices assume demand exceeds supply indefinitely. If the buildout creates sufficient supply, investors entering at peak valuations may not achieve the returns they expect.

Some projects will fail. At 1.1+ trillion dollars of announced investment, statistical certainty suggests some significant fraction will not succeed. The first major failures will test investor confidence and potentially reset valuations across the sector.

Sovereign wealth participation will grow. The largest projects require the longest time horizons and deepest capital pools. Their participation may become essential at the largest scales.

8.13 WHAT THE MONEY MEANS

David stands at his office window, thirty floors above Park Avenue, looking north toward Central Park. The late afternoon light catches something in the distance—a flash of gold among the bare December trees that might be the reservoir. Nature surrounded by concrete. In Michigan, concrete surrounded by nature. The city spreads below him, a hundred years of infrastructure investment made visible: subways, bridges, tunnels, towers. His reflection floats ghostlike in the glass—loosened tie, sleeves rolled to the forearm, the posture of a man who has been at his desk since 6 AM. Someone paid for all of it. Someone profited. Someone lost.

"You want to know if I think this is a bubble," he says, without turning from the window. His breath fogs the glass slightly. "The honest answer is I don't know. Nobody knows."

He has spent fifteen years in this business. He has seen cycles. The financial crisis, when he was young enough to think it couldn't happen again. The cloud boom. The crypto bust. Each time, smart people got it wrong. Smart money evaporated. He lost a college roommate to the 2008 crash—not dead, but broken, moved back to Ohio, never recovered. The market has no memory and no mercy.

"The difference this time is scale," he says. "The bets are bigger. The consequences are bigger. If we're right, we built the infrastructure for the next fifty years. If we're wrong..." He trails off. Below, a taxi honks. The sound barely reaches this height.

Frank sits in a township meeting room in Saline, Michigan. The walls are beige cinderblock, the same walls where he has sat through eight years of zoning variances and drainage disputes. The folding tables have seen better decades; someone has scratched initials into the laminate near his left hand. The fluorescent lights buzz faintly, one tube flickering in the corner. The room smells like old coffee and the industrial cleaner the janitorial service uses. Outside, through windows that need cleaning, he can see the construction site. Cranes against a gray sky. Earthmovers crawling like yellow insects. A preserved red barn that will soon be the quaintest thing for miles.

His hands rest on the table, palms down, the way they do when he is trying to steady himself. Thick hands, an engineer's hands, scarred from decades of projects in his garage. They built the deck where his grandchildren play. They cannot build anything that matters now.

He thinks about the investors he will never meet. The ones in Manhattan and Singapore and Abu Dhabi who made this their bet. He wonders what they see when they look at spreadsheets. Whether they picture the farms that used to be here. Whether they think about the families who will live with whatever comes next.

Probably not. Why would they?

The money flows from one place to another. It becomes concrete and steel, transformers and generators, cooling systems and fiber connections. It becomes the physical substrate of the digital economy. It becomes his township's future.

Frank did not choose this. Neither did his neighbors. The decision was made in rooms they will never enter, by people they will never meet, based on calculations they would not understand even if someone explained them.

But they will live with the consequences. For thirty years. For fifty. For however long the buildings stand. His granddaughter *Lily* will be forty-three when the first lease expires. She will have lived her entire adult life in server country.

The trillion-dollar buildout is not an abstraction. It is not a financial instrument or an investment thesis or a slide deck in a Manhattan conference room. It is *Harold*'s seven and a half million dollars—more than three generations of his family earned combined. It is *Ellen*'s retirement account, up twenty-three percent on the same companies paving over her neighbors' fields. It is *Frank*'s sleepless nights and *David*'s sleepless nights, different in every particular, alike in their uncertainty.

The facilities built today will stand for decades. They will either serve as essential infrastructure for an AI-powered economy or stand as monuments to a moment when investors believed something that turned out not to be true. No one can say which.

What we can say: the money is real. The land is real. The transformation is real.

The red barn in Saline Township, preserved for photographs, is real too. It will watch whatever comes next.

CHAPTER NINE

The Incentives

T HE PRESS CONFERENCE LOOKED LIKE EVERY OTHER ECONOMIC DE-
VELOPMENT ANNOUNCEMENT IN MICHIGAN HISTORY. Governor
Gretchen Whitmer stood at the podium, flanked by executives in dark
suits. Behind her, a banner displayed the state seal alongside corporate
logos. Cameras clicked. Reporters scribbled notes. Officials beamed with
the satisfied expressions of people about to claim credit for good news.

But the numbers were different. The scale was unprecedented.

On December 30, 2024, Whitmer signed Senate Bill 237 into law, creating Michigan's first targeted data center tax incentive program.[80] The
legislation exempted qualifying facilities from sales and use tax on equipment purchases—servers, cooling systems, generators, networking infrastructure, uninterruptible power supplies. For a typical hyperscale data
center spending billions on equipment, this exemption would be worth
hundreds of millions of dollars.

The ink had barely dried before the announcements began. Within
months, Michigan went from zero major data center investments to a
pipeline exceeding $11.5 billion.[18] Related Digital committed $7 billion to
Saline Township.[58] Form8tion announced $1 billion for Augusta Town-
ship. The University of Michigan partnered with Los Alamos National
Laboratory on a $1.2 billion AI research facility. Microsoft began quietly

purchasing land in Kent County. Whispers circulated about additional projects under negotiation, each larger than the last.

Michigan was not unique. Across America, states competed fiercely for data center investment, offering tax breaks, expedited permits, and infrastructure support worth billions. Virginia waived $928 million in sales tax revenue in fiscal year 2023 alone.[182] Georgia projected $327 million in foregone revenue for 2026. Kansas assembled a $128 billion pipeline through aggressive incentives and strategic positioning.

This chapter examines how that competition works: what states offer, why they offer it, who benefits, and who pays. We trace data center incentives from their origins in Virginia's internet boom to their current form as instruments of state industrial policy. We weigh the economic arguments for and against these programs, the evidence of their effectiveness, and the patterns that emerge from decades of experience. And we return to Michigan, where a law signed in December 2024 set the stage for everything that happened in Saline Township.

The incentive game reveals something essential about how data centers get built. Private capital flows toward public subsidy. The largest infrastructure investments in American history depend on tax exemptions negotiated behind closed doors. The communities that host these facilities often have the least say in the terms of their arrival.

9.1 Virginia's Advantage

The modern era of data center incentives began not with a policy decision but with a cable landing.

In the early 1990s, Northern Virginia became the hub of the commercial internet almost by accident. Government contractors clustered near the Pentagon and CIA had built fiber networks for classified communications. Those networks connected to MAE-East, one of the first major internet exchange points, housed in a nondescript building in Vienna. When the internet commercialized in the mid-1990s, the infrastructure was already in the ground. Decisions made by defense contractors in the 1980s created the topology that would define the internet for decades.

Companies like America Online, MCI/UUNET, and PSINet needed space for servers. They needed it close to the exchange points where data packets switched hands. They needed it fast—the internet was growing exponentially, and every month of delay meant lost opportunity. Northern Virginia had available office parks, empty warehouses, and a workforce comfortable with technical work. The data center industry emerged not from strategic planning but from proximity and timing.

The concentration accelerated through network effects. Each new data center made the location more attractive for the next. Interconnection between facilities grew easier and cheaper. Specialized vendors—fiber companies, cooling system manufacturers, security firms—clustered nearby. By 2000, Loudoun County alone hosted more data center capacity than most countries. The "Data Center Alley" stretching from Ashburn to Sterling became the densest concentration of digital infrastructure on Earth.

Amazon chose Northern Virginia for its cloud headquarters in 2006 not because of tax incentives but because of network topology. The fiber was in the ground. The substations were built. The workforce understood the technology. AWS US-East-1, launched from a facility in Ashburn, would become the largest cloud region in the world, processing more traffic than most nations' entire internet infrastructure.

Virginia's government noticed. Economic development officials saw tax revenue and jobs clustering in Loudoun and Prince William counties. But they also saw competition emerging. North Carolina was courting Google. Texas was promoting its deregulated power market. Arizona was highlighting cheap land and favorable climate. States that had paid little attention to data centers were beginning to offer incentive packages.

In 2009, facing this competition, the Virginia General Assembly passed the Data Center Retail Sales and Use Tax Exemption. The program exempted qualifying facilities from the 5.3 percent state sales tax on equipment purchases. For data centers spending hundreds of millions or billions on servers, storage, and networking gear, the savings ran to tens of millions of dollars.

The thresholds were modest by today's standards: $150 million in investment, fifty jobs paying at least 150 percent of the locality's average wage. These requirements reflected 2009 economics, when data centers cost tens or hundreds of millions of dollars and employed modest but well-paid technical staffs.

The program worked. Virginia maintained its dominance as other states developed competing packages. Investment kept flowing to Northern Virginia, and the state collected property taxes, income taxes, and utility taxes even as it waived sales taxes on equipment.

But the program worked perhaps too well. By 2023, Virginia was waiving $928 million annually in sales tax revenue.[182] Northern Virginia hosted 13 percent of global data center capacity—a remarkable concentration for a single region. The exemption had become so embedded in industry expectations that removing it would be economically suicidal. Other states had copied the model, creating a baseline that Virginia had to maintain just to stay competitive.

Virginia's experience established the template that would shape data center policy nationwide. First, infrastructure attracted investment organically—the internet's topology created natural clustering points. Then, when other states began competing, incentives became necessary to maintain position. Finally, the incentives grew so large and so expected that their removal became politically impossible. A tax break that began as a competitive tool became an ongoing subsidy to an industry that had long since established itself.

The pattern would repeat, with variations, in state after state. The incentives grew larger. The competition intensified. And the fundamental question—whether tax exemptions for data centers represent good public policy—became increasingly difficult to answer.

9.2 ANATOMY OF INCENTIVES

Data center incentives come in several forms, often stacked together in packages worth billions. Understanding these mechanisms is essential to understanding how the industry operates and why projects land where they do.

The most common incentive is exemption from state and local sales tax on qualifying purchases. A hyperscale facility might spend $4 billion on equipment during its initial buildout, with ongoing replacement and expansion adding billions more over its lifetime.

At a typical 6 percent sales tax rate, a $4 billion equipment purchase represents $240 million in tax liability. Exempting this purchase creates an immediate, quantifiable benefit. The savings can be calculated precisely and incorporated into financial models. They reduce upfront capital requirements and improve return on investment.

Eight of the top ten data center states offer some form of sales tax exemption. The terms vary significantly. Virginia requires $150 million in investment and fifty jobs paying at least 150 percent of the locality's average wage. Georgia's thresholds range from $100 to $250 million depending on county population, plus twenty-five "quality jobs" paying 110 percent of the county average. Illinois requires $250 million and twenty jobs. Texas requires only twenty jobs for investments over $200 million— one job per $10 million invested. Nevada requires $25 to $100 million depending on jurisdiction, with job thresholds as low as ten positions.

The job requirements reveal an awkward truth. Data centers are capital-intensive, not labor-intensive. A $1 billion facility might employ only fifty to two hundred permanent workers once construction ends. A

$1 billion manufacturing plant might employ five hundred to two thousand. A $1 billion hospital might employ three thousand or more.

States must set job thresholds low enough that data centers can realistically qualify. Set requirements too high, and you disqualify the very investments the program is designed to attract. Set them too low, and the foregone revenue looks hard to justify. This tension between job creation rhetoric and capital-intensive reality runs through every data center incentive program.

———

Local governments control property taxes, and they often negotiate abatements or reductions for major data center investments. These arrangements take several forms, each with distinct implications for local budgets.

Some localities reduce the assessment rate for data center equipment. Henrico County, Virginia, cut its data center property tax rate from $3.50 to $0.40 per $100 of assessed value in 2017—an 89 percent reduction. The county later raised the rate to $2.60 per $100 in 2025, reflecting revised calculations about fiscal impact. The reduction applied specifically to data center personal property, leaving real property taxed at normal rates. The structure preserved some revenue while dramatically reducing the tax burden on capital-intensive equipment.

Arizona allows accelerated depreciation that reduces taxable assessment by 75 percent in the first year, phasing out over subsequent years. This front-loads benefits for developers facing high costs during initial construction, then gradually increases the tax burden as facilities mature and begin generating revenue.

Other localities use Payment in Lieu of Taxes (PILOT) agreements, where the data center pays a fixed annual amount rather than taxes based on assessed value. These agreements provide budget certainty for local governments—they know exactly what they will receive regardless of how the project's value changes. They also cap liability for developers, who can model costs precisely without worrying about assessment increases.

A typical PILOT for a hyperscale facility might specify $10 million annually for twenty years, regardless of the facility's assessed value. If the facility would otherwise generate $50 million in property taxes, the PILOT represents an 80 percent reduction. If assessment methodologies would produce only $15 million in taxes, the PILOT provides budget stability for the locality while still reducing the developer's burden.

Nevada offers a particularly aggressive structure: 75 percent abatement on personal property taxes for ten to twenty years, depending on investment scale. For a facility with $2 billion in equipment, this could mean hundreds of millions in savings over the abatement period. The structure helped Nevada attract data center investment despite lacking the fiber connectivity and existing clusters that benefited Virginia and Texas.

———

Some incentives take the form of direct government investment in enabling infrastructure. Roads, water systems, electrical substations, fiber connectivity, wastewater treatment—these cost money that developers would otherwise provide themselves. When governments fund infrastructure, they reduce project costs while potentially providing benefits that extend beyond any single facility.

When Meta announced its $10 billion data center in Richland Parish, Louisiana, the state committed over $200 million in infrastructure.[183] State Route 134 would be widened to handle construction traffic and permanent operations. Water treatment capacity would expand. Electrical transmission lines would be upgraded. Natural gas pipelines would be extended. The state would fund improvements to roads, bridges, and utilities throughout the region.

237

The arrangement had elegant political logic. State money flowed to Meta indirectly, through infrastructure the public would also use. The expenditure appeared as transportation and utility investment rather than corporate subsidy. Press releases emphasized community benefits—better roads, improved water service, enhanced emergency response—without dwelling on the fact that Meta's project necessitated the spending.

For the community, the calculus grew complicated. Some improvements would benefit residents for decades. Others were sized specifically for data center needs and would provide minimal public benefit. A substation capable of serving 500 megawatts serves the data center; the surrounding community might need 10 megawatts. The public funded infrastructure primarily designed for private use.

Kansas took infrastructure support further through its "super project" designation for investments exceeding $1 billion. Qualifying projects receive expedited permitting—ninety days maximum for all required approvals—plus dedicated state assistance navigating regulatory processes. The state assigns a concierge team to shepherd projects through bureaucratic requirements that might otherwise take years.

Electricity is the largest ongoing cost for data centers. Industrial rate structures, volume discounts, and special tariffs can reduce this cost substantially. Cumulative savings over a facility's lifetime dwarf upfront tax exemptions.

Consider the economics. A 500-megawatt data center running at 90 percent utilization consumes roughly 4 billion kilowatt-hours annually. At typical commercial rates of $0.10 per kilowatt-hour, that means $400 million per year in electricity costs. A 30 percent discount reduces the annual bill to $280 million—$120 million in savings, every year, for the facility's entire operating life.

In Texas, the deregulated electricity market allows data centers to contract directly with generators, often securing rates 30 to 40 percent below what residential customers pay. Large users can negotiate power purchase agreements with wind farms, solar installations, or natural gas generators, locking in prices for years or decades. Cheap power and contractual certainty have made Texas the fastest-growing data center market in America.

California's PG&E Direct Access program offers qualifying facilities rates around 9 cents per kilowatt-hour compared to standard commercial rates of 13 to 15 cents. The discount reflects the value utilities place on large, predictable, creditworthy customers. Data centers consume electricity steadily, twenty-four hours a day, every day of the year. They do not create the demand spikes that challenge grid management. For utilities, they are ideal customers.

These discounts spark controversy. When utilities spread infrastructure costs across all customers, residential ratepayers subsidize large industrial users. A new substation built to serve a data center might cost $50 million, funded through rate increases for everyone even though only the data center benefits directly. The data center pays discounted rates. Everyone else pays more.

This cost-socialization concern has driven recent legislative action in several states, a topic we explore in detail later in this chapter.

Some states offer workforce development incentives targeting data center employment. These programs fund training for data center technicians, provide wage subsidies for new hires, or support educational programs at community colleges and technical schools.

Georgia's Quick Start program provides customized training for data center employees at no cost to the employer. The state funds curriculum development, instructor salaries, equipment, and facilities. Workers receive training. Employers receive a skilled workforce. The state hopes to attract investment that would otherwise go elsewhere.

Virginia's Virginia Jobs Investment Program offers per-employee subsidies for data center hiring, particularly for positions filled from the unemployment rolls or from populations facing barriers to employment. A hyperscale facility hiring two hundred workers might receive $1 million or more in training grants.

The workforce development approach addresses a genuine constraint. Data centers require specialized skills that traditional education does not provide. Cooling system technicians, electrical engineers, security specialists, and facilities managers must understand technologies that change rapidly. Training programs accelerate workforce development and reduce hiring costs.

Critics note that workforce subsidies transfer public funds to highly profitable companies that could afford to train their own employees. The counterargument: subsidized training benefits workers who might otherwise lack access to well-paying technical careers.

9.3 STATE PROFILES

Examining specific states reveals strategic choices: how to attract investment, and who bears the costs.

Virginia faces what economists call the incumbent's dilemma. The state pioneered data center incentives in 2009, attracted over 300 operational facilities with combined capacity exceeding 13 gigawatts, and established the sales tax exemption template that other states now match. But dominance has eroded. Virginia's share of new U.S. investment dropped from roughly 40 percent in 2020 to about 25 percent in 2025. Other states matched the incentives while offering cheaper land and available power. Loudoun County, which hosts the densest concentration of data centers on Earth, has begun pushing back with zoning restrictions. Virginia must spend ever more to maintain position, even as returns diminish.

Texas competes on power and speed rather than tax incentives. The ERCOT grid allows data centers to contract directly with generators, securing rates 30 to 40 percent below regulated markets. ERCOT's interconnection process can complete in months; PJM's takes years. For companies racing to deploy AI training infrastructure, that timeline difference matters more than any tax exemption. The risks are equally significant. Winter Storm Uri in 2021 killed more than 200 people when the isolated grid nearly collapsed.[85] But speed and cost trump resilience concerns for many operators.

Kansas shows how aggressive incentives can transform a state's position almost overnight. In 2022, the state had no hyperscale presence. By December 2025, it had assembled a $128.8 billion pipeline across nine major projects, including Project Kestrel ($100 billion)[135] and Red Wolf DCD Properties ($12 billion).[184] The state offered 100 percent sales tax exemption, property tax abatements up to thirty years, and expedited permitting capped at ninety days. Whether the $128 billion pipeline translates to completed facilities remains uncertain. The foregone revenue, if it does, will be substantial.

Pennsylvania highlights the limits of brownfield incentives. The state's coal, steel, and manufacturing heritage left thousands of abandoned industrial sites, many with existing electrical infrastructure. Policy logic favors reusing these sites over consuming farmland. In practice, projects like Homer City (4.5 GW) and TECfusions Keystone (3.0 GW) face environmental contamination, complex title issues, and remediation delays that stretch timelines beyond greenfield alternatives. Speed beats legacy site reclamation—a pattern explored in Chapter 7.

9.4 THE RACE TO THE BOTTOM

State competition for data center investment has intensified dramatically since 2020. The dynamics resemble classic economic development races, but the scale and speed are without modern parallel.

When Virginia offered sales tax exemptions in 2009, other states followed. When Georgia added property tax abatements, competitors matched. When Texas promoted cheap power, Arizona developed similar pitches. The race has no obvious endpoint. Investment thresholds rise. Job requirements fall. Exemption periods lengthen. The competitive position that incentives were supposed to create erodes as everyone offers the same deal.

The race to the bottom raises a fundamental question: if practically every state offers equivalent incentives, do the incentives affect investment decisions at all? Academic research is mixed. Some studies find incentives influence location decisions. Others find they primarily affect timing. Still others find minimal impact, with location decisions driven by infrastructure and market factors that incentives cannot override. States continue offering incentives partly because they cannot risk being the first to stop. Prisoner's dilemma logic drives policy even when participants recognize its problems.

David has seen the incentive game from the capital side. His fund has lobbied for state incentives in Virginia, Georgia, and Texas. They contributed to industry associations that advocate for favorable tax treatment. They hired consultants who knew which state officials to call.

"I am not proud of all of it," *David* admits. "We are very good at getting states to compete against each other. That is literally our job—to extract the best terms possible. But when you see what some of these communities give up..." He trails off. His fund's investors—pension funds, university endowments, insurance companies—expect returns. The incentives improve those returns. The logic is straightforward even when the ethics are complicated.

He has watched ratepayer protection debates unfold in state capitals. He understands why residents in Virginia and Michigan worry about subsidizing data center infrastructure. He also knows his fund benefits from socialized costs. "The system works for us," he says. "I am not sure it works for everyone else."

Export policy adds another layer of complexity. "When the H200 approval came through in December, some of my partners were thrilled," *David* says. "NVIDIA's stock recovered. Their margins improve. But think about what it means for domestic infrastructure investment. If Chinese customers can buy H200s legally, some workloads that might have run on American facilities now run on Chinese ones instead. We're competing with our own chip exports."

He pauses, running through the implications. "The revenue share captures something for the government. It does not capture the value of domestic data center jobs, or the infrastructure investment that would have happened here. Policy optimized for chip sales is not the same as policy optimized for infrastructure investment. Sometimes they conflict."

Frank understood township zoning. He had spent fifteen years on the board, reviewing proposals for subdivisions and strip malls. He could read a site plan, identify drainage issues, ask the right questions about setbacks and parking ratios. Then Related Digital arrived with a $7 billion proposal, 1.4 gigawatts of power demand, and engineering specifications running to thousands of pages.[185] GPU thermal envelopes, interconnection queue positions, cooling system PUE ratings—he did not know what half the terms meant.

He spent a weekend trying to understand the technical documents. His engineering background helped with some of it. But the scale was incomprehensible. One point four gigawatts. He looked it up: enough power for a city of 800,000 people. Saline Township had 2,300.

"How am I supposed to make a decision about something this big?" he asked his wife one evening, the documents spread across the dining room table. She had no answer. Neither did the other board members when he asked them. They were a retired insurance agent, a part-time farmer, a woman who ran the township's historical society, a man who fixed small engines. Good people. Intelligent people. People completely unprepared for what they were being asked to decide.

The science and technology scholar Sheila Jasanoff calls this a "democracy deficit"—the gap between the decisions citizens must make and the knowledge required to make them well.[186] Township boards approve or deny based on trust, not analysis. State legislators vote on incentive programs based on economic projections they cannot verify. The complexity of AI infrastructure systematically exceeds the analytical capacity of democratic institutions.

Michigan's SB 237 passed after testimony from executives, economists, and utility representatives—all with interests in particular outcomes, all possessing information asymmetries the legislature could not overcome. By the time anyone could evaluate whether it worked, the legislators who voted for it would be long out of office.

9.5 MICHIGAN'S ENTRY

Frank watched the legislative session streaming from Lansing on his laptop, sitting at his kitchen table in Saline Township, sixty-five miles from the capitol. October 2024. Senate Bill 237 was working through committee. Related Digital had already been buying options on farmland—he knew because *Harold* had called him, voice uncertain, asking what he should do about the men in suits who kept raising their offers.

DTE Energy had initiated interconnection studies. The pieces were in motion, waiting for the policy framework to lock into place.

An economic development official described the potential: billions in investment, thousands of jobs, tax revenue that would transform rural communities. The official was young, articulate, confident. He used words like "opportunity" and "partnership" and "future." He did not mention what would be given up. He did not mention the farms. He did not mention the communities that would be transformed whether they wanted transformation or not.

Frank's phone buzzed. A text from *Harold*: "They make it sound so simple. Like we're just numbers on a spreadsheet."

Frank called him back. Forty years of sharing fence lines and equipment, of showing up when someone's machinery broke down. That kind of neighbor.

"What are you going to do?" he asked.

"I don't know." *Harold*'s voice was tired. "They're offering three times what the land is worth. My kids don't want to farm it. Ellen's boys don't want to farm it. Nobody's kids want to farm anything anymore."

"Ellen's holding out."

"Ellen's stubborn. Always has been." *Harold* paused. "Maybe she's right."

Frank thought about what he should say. As township supervisor, he had an interest in the outcome—the project would transform his community, for better or worse. As *Harold*'s neighbor, he had a different kind of stake.

"I can't tell you what to do," *Frank* said finally. "Nobody can. But whatever you decide, Harold—I want you to know I understand. Either way, I understand."

There was a long silence on the line. When *Harold* spoke again, his voice was rough. "That means something. Thank you."

That was the problem, *Frank* thought after he hung up. To legislators in Lansing, to executives in California, to the young official with the confident voice, they were numbers. Megawatts and dollars. Tax revenue and job projections. The people who would live with the consequences were not in the room. They were never in the room.

By December, he knew, the bill would become law. And *Harold* would make his decision, and everything that followed would flow from legislators voting on a bill they did not fully understand for communities they had never visited.

Michigan entered the data center incentive competition late. The state had data centers—colocation facilities in the Detroit suburbs, regional operations for various enterprises, disaster recovery sites for financial institutions—but nothing at hyperscale. The major investments went elsewhere: Virginia, Texas, Arizona, Georgia.

Governor Whitmer's administration identified data centers as a strategic priority during her second term. Michigan had advantages: reliable power from DTE Energy and Consumers Energy with substantial available capacity, affordable land outside the Detroit metropolitan area, excellent fiber connectivity through existing network infrastructure, and a skilled workforce trained by the automotive and technology industries.

What Michigan lacked was a competitive tax structure. Without exemptions comparable to Virginia or Georgia, Michigan could not compete for hyperscale investment. Site selection consultants would not include Michigan on their short lists. The state was not in the game.

Senate Bill 237, introduced in October 2024, addressed this gap. The legislation created a new sales and use tax exemption for qualifying data center equipment. To qualify, a facility needed to meet investment thresholds, create a minimum number of jobs paying specified wages, and locate in an eligible zone designated by the Michigan Economic Development Corporation.

The bill's structure borrowed from successful programs in other states while adapting to Michigan's circumstances. Investment thresholds matched Georgia and Virginia. Job requirements reflected data center economics—modest numbers, high wages. The MEDC's role in designating eligible zones gave the state power to direct investment toward preferred locations.

The bill moved quickly through the legislature. Economic development officials testified about job creation and investment potential. Utility executives described available grid capacity. Technology industry representatives explained why Michigan's workforce and infrastructure made it attractive. The testimony created an impression of broad support and urgent need.

Critics raised concerns that received less attention. Environmental groups warned about energy consumption and carbon emissions. Consumer advocates questioned whether residential ratepayers would bear grid upgrade costs. Transparency advocates noted that many deal terms would be negotiated privately, without public disclosure requirements. Local government representatives asked whether townships and counties would have meaningful input into projects transforming their communities.

The concerns did not derail the legislation. On December 30, 2024, Governor Whitmer signed the bill into law at a ceremony featuring executives from multiple technology companies expressing interest in Michigan. The event was designed to demonstrate momentum, to signal that Michigan was open for business, to create urgency before the session ended.

Within weeks, announced investments exceeded $11 billion. Related Digital committed $7 billion to Saline Township. Form8tion announced $1 billion for Augusta Township. The University of Michigan partnered with Los Alamos National Laboratory. Microsoft began purchasing land.

Senate Bill 237 did not create the Saline Township project. Related Digital had already begun assembling land and negotiating with DTE Energy before the law passed. The company's agents had purchased options on farmland parcels throughout 2024. DTE had initiated interconnection studies. The project was underway before incentives existed.

But the legislation created the framework that made Michigan competitive. It signaled to the industry that the state was serious about attracting investment. It removed a barrier that had kept Michigan off site selection short lists. And it provided a political backdrop against which specific projects could be announced and celebrated.

Related Digital was well-positioned to navigate that backdrop. Stephen Ross had employed the current Secretary of State at his philanthropic foundation before she entered politics; her husband worked as a vice president at Related Companies.[187] None of this was improper. But it illustrated how major deals in a small state flow through networks where the same names appear in multiple roles.

The relationship between incentive and investment is rarely simple. Companies do not decide where to build based solely on tax policy. Power availability, land costs, workforce quality, regulatory environment, fiber connectivity, and dozens of other factors matter. But incentives can tip the balance when multiple sites are otherwise comparable. They send a signal about state priorities that influences how companies perceive their options.

Michigan's signal was clear. The state wanted data centers. It was willing to compete with Virginia, Georgia, and Texas. And it would offer the tax treatment necessary to attract major investments.

Michigan's 2024 data center legislation built on two decades of incentive policy. The state had learned hard lessons. Film subsidies recovered only eleven cents per dollar of credits granted.[188] The MEGA credit program created billions in long-term obligations.[189,190] Both taught Michigan that open-ended tax credits could create fiscal burdens outlasting the conditions that justified them. Senate Bill 237 reflected those lessons, borrowing from other states while adding accountability mechanisms.

9.6 THE JOBS QUESTION

Every data center announcement includes job numbers. Governors stand at podiums and cite employment figures. Press releases trumpet job creation. Economic impact analyses project multiplier effects. Everyone celebrates.

The numbers deserve scrutiny.

Data centers create few permanent jobs per dollar invested. A typical hyperscale facility employing one hundred workers might require $2 billion in capital—one job per $20 million. Manufacturing typically creates one job per $1 to $5 million. Hospitals create one job per $500,000 to $1 million. Retail creates one job per $100,000 to $300,000.

The Saline Township project illustrates this starkly. Related Digital's $7 billion investment would create approximately 450 permanent jobs according to project documentation.[57] That works out to roughly $15.5 million per job—better than average for hyperscale data centers but far below any manufacturing benchmark.

For context: the same investment in automobile manufacturing might create 5,000 to 10,000 permanent jobs, in healthcare infrastructure 15,000 to 20,000, in education even more. Data centers are capital-intensive in ways few other industries match.

Construction jobs present a different picture. Building a $7 billion facility requires thousands of workers over several years. Related Digital estimated 2,500 union construction jobs during the buildout period. These are real jobs paying real wages, and they matter to the workers who hold them. The construction trades have become strong supporters of data center development precisely because the sector provides substantial work.

But construction jobs end. When the facility opens, the workforce moves on. The permanent staff is modest. The wages are high—$80,000 to $150,000 for technical positions—but they benefit only a small number of workers.

This creates an awkward political dynamic. States justify incentive programs based on job creation, but data centers are not job-creating engines in the traditional sense. They are capital-intensive facilities that employ relatively few people to operate vast amounts of equipment. The rhetoric emphasizes jobs. The economics deliver capital.

Proponents offer several responses to this criticism.

First, they argue that the jobs are high-quality positions requiring specialized skills and paying well above median wages. Data center technicians earn two to three times what retail or food service workers earn. Quality matters as much as quantity. A hundred jobs at $100,000 per year may provide more community benefit than five hundred jobs at $30,000.

Second, they point to indirect employment. Construction requires materials, which require manufacturing, which requires workers. Operations require services—security, janitorial, food service, transportation, maintenance—that create additional jobs. Utility expansion creates employment for lineworkers and engineers. The economic ripple effects extend beyond direct data center employment.

Third, they emphasize tax revenue. Even with generous exemptions, data centers pay substantial property taxes, income taxes from highly paid workers, and various fees. A $7 billion facility might generate $50 million or more annually in local property taxes, depending on assessment and abatement terms. That revenue funds schools, roads, and public services benefiting everyone.

Fourth, they argue for cluster effects. Data center clusters attract related businesses: chip designers, cooling system manufacturers, fiber network operators, cloud service providers, software firms. Northern Virginia's data center concentration helped attract Amazon's second headquarters, bringing tens of thousands of additional jobs. Cluster effects may exceed direct employment impact.

Critics find these arguments insufficient. Job-per-dollar ratios matter because state resources are finite. Money spent on data center incentives cannot fund other economic development that might create more jobs. The opportunity cost must be considered.

They also question the multiplier estimates that economic impact studies typically cite. Many such analyses come from consultants hired by developers or economic development agencies. These analysts have incentives—financial and professional—to present favorable findings.

Their models often use assumptions that inflate projected benefits. Independent verification is rare.

The debate has no clear resolution. Data centers create jobs, but fewer than their investment size suggests. The jobs are high-quality, but they benefit relatively few workers. The economic effects are real but hard to quantify. States must make policy decisions based on incomplete information and competing claims.

9.7 RATEPAYER PROTECTION

Cost socialization has become the most contentious issue in data center policy. When utilities build infrastructure to serve data centers, who pays?

Traditionally, utilities spread infrastructure costs across their entire customer base through rate increases. A new substation serving a data center might cost $50 million. That cost gets added to the utility's rate base and recovered through higher bills for all customers—residential, commercial, and industrial alike.

This approach made sense when large customers were rare and their infrastructure needs modest. A new factory might require a small substation upgrade. The cost, divided among millions of ratepayers, was negligible. The factory provided economic benefits—jobs, tax revenue, supply chain purchases—that justified socialized infrastructure investment.

The approach breaks down when single customers demand gigawatts of power and require billions in grid upgrades. The scale has changed. The regulatory framework has not.

Consider the Saline Township project. The facility needs 1.4 gigawatts of power. Delivering that power requires new transmission lines, expanded substations, upgraded distribution infrastructure, and possibly new generating capacity. The costs could reach hundreds of millions of dollars.

If these costs are spread across all DTE Energy customers, residential ratepayers subsidize OpenAI's electricity consumption. A retired schoolteacher in Dearborn pays higher electric bills so that an artificial intelligence company valued at over $100 billion can train language models. The distributional implications are troubling.

States have begun addressing this through what advocates call "cost-responsibility" models. The core idea is simple: data centers should pay for the infrastructure they require rather than passing those costs to other customers.

Texas led with Senate Bill 6, signed into law in June 2025.[91,191] The law requires large-load customers—those demanding more than 75 megawatts—to pay their own grid connection costs. They must demonstrate financial ability to fund necessary upgrades. They must also provide backup power equal to 50 percent of their contracted load, protecting the ERCOT grid during peak demand or system stress. ERCOT, the Electric Reliability Council of Texas, operates largely isolated from the rest of the country, making grid stability a state-level concern.

Oregon followed with the POWER Act in 2025.[192] Facilities using more than 20 megawatts must enter minimum ten-year contracts guaranteeing payment. They must sign power purchase agreements demonstrating how they will meet their load. They must pay their fair share of grid expansion rather than passing costs to residential customers.

Pennsylvania is developing similar requirements. The state Public Utility Commission began proceedings in 2025 to establish a model tariff for large-load customers. The goal: a framework that protects residential ratepayers from subsidizing data center infrastructure while still accommodating industry growth.

Georgia took a different approach with its 100-megawatt rule. Utilities must demonstrate that large data center loads will not increase rates for existing residential customers. Georgia Power must show regulators that infrastructure investments are cost-neutral or beneficial to the broader customer base. If analysis suggests rate increases would result, approval can be denied.

Michigan faced this question during the Saline Township controversy. Attorney General Dana Nessel filed a brief urging the Michigan Public Service Commission to scrutinize DTE Energy's request for expedited approval of special power contracts.[93] Residential ratepayers deserved protection from potential cost increases, she argued, and the commission should not rush approval without thorough analysis.

DTE Energy responded that the data center would actually decrease residential rates by spreading fixed grid costs over greater electricity sales. More electricity sold means more revenue to cover infrastructure, reducing the per-customer burden. The company presented modeling claiming $300 million in net benefits to other ratepayers over the contract term.

This argument has superficial appeal but ignores important dynamics. If the data center requires $500 million in new infrastructure, spreading that cost over more customers only helps if the new revenue exceeds the new costs. If the data center receives discounted rates—as industrial customers typically do—the revenue may not cover the infrastructure investment.

The modeling matters, but the modeling was not publicly available. Key provisions of DTE's contract with Related Digital were redacted as proprietary business information. The attorney general noted that verifying the utility's claims was "impossible" without access to actual contract terms.[7]

The cost-responsibility movement reflects a broader political shift. For decades, economic development agencies focused on attracting investment with minimal conditions. Benefits were assumed to flow automatically. Now, legislators and regulators ask harder questions about who benefits and who pays.

9.8 THE FEDERAL ROLE

State incentives do not operate in isolation. Federal policy shapes the environment in ways that amplify—and sometimes distort—state competition.

The Inflation Reduction Act of 2022 transformed data center energy economics. Clean energy provisions—production tax credits, investment tax credits, domestic content bonuses—make renewable energy dramatically cheaper. A solar installation qualifying for the full IRA stack might receive federal support equal to 40 to 50 percent of project costs. If Washington subsidizes renewable energy, states can focus limited resources on other dimensions of competition.

The CHIPS and Science Act has similar effects.[193] By pouring $52 billion into semiconductor manufacturing, the law increases demand for data center capacity near chip fabrication plants. Arizona's data center boom cannot be separated from TSMC's $65 billion investment in Phoenix.[194] Intel's $20 billion fab in Ohio drives data center development in Columbus.[195] Federal semiconductor policy creates data center demand; state incentives capture it.

Accelerated depreciation for equipment allows data centers to deduct costs over five to seven years even though equipment may operate for fifteen or twenty. These benefits apply regardless of location, moderating the impact of state incentive differences. The likely trajectory: continued federal subsidy through energy and technology programs, with states competing at the margin for investment that federal policy enables.

9.9 WINNERS AND LOSERS

The incentive game produces clear winners. Technology companies receive billions in tax savings that flow to their bottom lines. Developers secure projects that might otherwise go elsewhere. Utilities gain customers who consume vast quantities of electricity at predictable rates. Construction workers find employment. Landowners receive premium prices for agricultural parcels. Economic development officials claim credit for investment announcements. Politicians celebrate ribbon cuttings.

The losers are harder to identify but no less real.

Taxpayers who do not work for data centers pay the foregone revenue indirectly. Schools, roads, and public services that might have been funded by data center taxes go unfunded or require higher taxes on everyone else. The $928 million Virginia waives annually represents teachers not hired, roads not paved, social services not provided. The money did not disappear. It went to corporate treasuries.

Ratepayers may face higher utility bills. If infrastructure costs are socialized, residential customers subsidize industrial operations. The subsidy is invisible—it appears as a rate increase rather than a transfer payment—but it transfers value from households to corporations.

Communities that host data centers bear costs that incentive packages do not address. Traffic increases during construction years. Noise from cooling systems runs continuously. Water resources face new demands. Agricultural land disappears under server halls. Rural character transforms in ways residents did not choose and may not welcome.

Workers in other industries receive smaller incentives per dollar invested. States have limited resources. Money spent attracting data centers cannot be spent attracting manufacturers, research facilities, hospitals, or other employers with higher job intensity. The opportunity cost falls on industries that states choose not to prioritize.

Future taxpayers inherit the consequences of current decisions. Many incentive agreements extend twenty or thirty years. The politicians who negotiate them will be long gone when the full costs become apparent. Benefits are immediate and visible. Costs are deferred and diffuse.

This distribution raises questions about democratic accountability. Most incentive negotiations happen in private. Deal terms are often confidential, shielded by trade secret exemptions from open records laws. Citizens learn the results only after agreements are signed. Communities most affected often have the least say in the terms.

Saline Township illustrates the problem. Residents voted through their township board to reject the project. The developer sued.[1] The township, facing legal costs it could not afford, settled and reversed its decision.[56] The project proceeded over the objections of people who had exercised their democratic voice and found it did not matter.

The incentive system did not cause this outcome directly. But it created the conditions that made it possible. Billions of dollars in potential investment justified aggressive legal action. The state's welcoming posture signaled that local resistance would not be tolerated. Federal officials celebrated the project as a national strategic asset. Against these aligned interests, a township of 2,300 people had little bargaining power.

9.10 TRANSPARENCY

The lack of transparency in incentive negotiations draws criticism from advocates across the political spectrum. Deals worth billions are negotiated behind closed doors. Terms are confidential, protected by trade secret exemptions. Economic projections come from interested parties. Independent analysis is rare. Citizens must trust that officials negotiated well.

Virginia's Joint Legislative Audit and Review Commission began reviewing the state's data center incentive programs in 2024. The investigation examined whether the annual foregone revenue generates commensurate economic benefits—and whether incentives were necessary to attract investment that would have come anyway.

The questions are harder than they seem. No one knows the counterfactual—what would have happened without incentives. Would Amazon have built AWS US-East-1 in Northern Virginia without tax exemptions? Probably, given the fiber infrastructure already in place.

But would Meta's latest campus have chosen Virginia over Georgia without incentives? Possibly not.

The attribution problem makes evaluation difficult. Investment follows many factors. Separating the effect of incentives from infrastructure, workforce, location, and market conditions requires assumptions that reasonable people dispute.

Georgia has developed accountability mechanisms worth examining. The state requires surety bonds of up to $20 million to ensure compliance with incentive terms. Companies that fail to meet investment or job commitments forfeit the bond. Annual compliance reporting is mandatory. The state maintains ongoing control rather than giving away benefits upfront.

Illinois takes a different approach. The state's twenty-year sales tax exemption is structured as five-year certificates that companies must renew by demonstrating continued compliance. This maintains state oversight throughout the incentive period rather than granting irrevocable benefits. Companies that close facilities, reduce employment, or fail to meet commitments can lose their exemptions.

Oregon's Strategic Investment Program requires qualifying projects to pay a community investment fee equal to 25 percent of abated taxes or $2.5 million annually, whichever is greater. This ensures that even exempt facilities contribute to the communities where they operate. The fee funds local services, infrastructure, and workforce development—benefits that flow to residents regardless of whether they work for the data center.

These accountability measures represent progress but remain incomplete. Enforcement is inconsistent; violations are often resolved through negotiation rather than clawback. Economic development agencies have incentives to report success rather than scrutinize failures. Confidentiality requirements limit public understanding of what deals actually contain.

The movement toward transparency is growing. Advocacy groups like Good Jobs First publish analyses of incentive programs. Investigative journalists examine specific deals. Academics study the economic effects of different structures. State legislators ask harder questions before approving new programs.

But the fundamental tension remains. Incentive competition creates pressure to offer generous terms and close deals quickly. Transparency slows the process and reveals the extent of public subsidy. Democratic accountability and competitive positioning pull in opposite directions, with no easy resolution.

9.11 Lessons Learned

What can we learn from the experience of data center incentives across America?

First, incentives matter less than infrastructure. Virginia became a data center hub because the internet's backbone ran through Northern Virginia. Texas attracts investment because ERCOT provides cheap, deregulated power. Arizona benefits from semiconductor industry clustering. States without these fundamentals cannot compete their way to success through tax policy alone. Incentives can tip the balance between comparable sites. They cannot overcome fundamental deficiencies.

Second, first-mover advantage is real but temporary. Virginia's early dominance has eroded as other states matched its incentives and offered additional advantages. Competitive position must be constantly renewed as rivals catch up. Early success does not guarantee continued success.

Third, job creation claims should be viewed skeptically. Data centers are not jobs programs. They are capital-intensive facilities that employ few workers per dollar invested. Economic justifications based on job creation often overstate employment effects while understating costs. Honest discussion requires acknowledging job creation's limits as justification.

Fourth, cost-responsibility frameworks are becoming standard. The era of unlimited cost socialization is ending. States increasingly require data centers to pay for infrastructure they require rather than passing costs to residential ratepayers. This trend will continue as utilities and regulators recognize the scale of infrastructure demands.

Fifth, transparency improves outcomes. States with strong reporting requirements and accountability mechanisms tend to get better deals. Confidentiality benefits developers at the expense of public interest. Greater transparency should be encouraged.

Sixth, local communities need more power. The pattern of local rejection followed by legal pressure and state override serves no one well. Communities should have meaningful input into developments that will transform them, even when those developments bring economic benefits. The current system prioritizes state and corporate interests over local self-determination.

These lessons suggest directions for reform. States could coordinate to reduce competitive pressure, though such coordination faces collective action problems. Federal standards could establish minimum taxation levels, though federal intervention in state economic development would be controversial. Local governments could receive stronger legal protections against override, though such protections would complicate state strategies. Transparency requirements could ensure public understanding of deal terms, though developers would resist disclosure.

Whether reforms will occur remains uncertain. The interests benefiting from the current system are powerful and well-organized. The interests harmed are diffuse and poorly represented. Political economy favors continuation of existing patterns, even as those patterns generate increasing criticism.

9.12 THE SALINE TOWNSHIP CONNECTION

Every element of this chapter connects to what happened in Saline Township.

Michigan's Senate Bill 237 created the tax framework that made the project economically attractive. The law's passage in December 2024 signaled that Michigan was competing seriously for data center investment. Without the exemption, Related Digital might have gone elsewhere.

The race to the bottom meant Related Digital had alternatives. If Michigan did not offer competitive terms, Virginia or Georgia or Kansas would. The threat was never explicit but always present. The company's negotiating position drew strength from the knowledge that investment is mobile and states are hungry.

The job creation claims followed the industry pattern: 450 permanent jobs and 2,500 construction jobs. Real jobs supporting real families. But they came with $7 billion in investment—a ratio that makes sense only if job creation is not the primary justification.

The ratepayer protection controversy played out in MPSC hearings. DTE Energy sought expedited approval for special power contracts. Environmental groups and consumer advocates demanded scrutiny. The attorney general urged caution. The commission faced pressure from state officials celebrating the economic development win to approve quickly.

The transparency concerns manifested throughout the project's history. Related Digital assembled land without revealing who would occupy the facility. Deal terms emerged only after agreements were finalized. Key provisions of power contracts were redacted. Citizens learned the full scope of what was planned for their community only after they lost the power to stop it.

The local community's loss of control defined the trajectory. Saline Township voted no. The developer sued. The township, outmatched in legal resources, settled. The project proceeded. Democratic voice was heard—and overruled by economic power.

These patterns did not originate with Saline Township. They represent how data center development works across America. The incentive system creates them. Competitive dynamics sustain them. Political economy makes them difficult to change.

Understanding incentives is essential to understanding why data centers go where they go and what happens to communities when they arrive. The money flows in particular directions for particular reasons. Those reasons are not mysterious. They are simply not discussed in the press releases and podium speeches that accompany economic development announcements.

The student in Ann Arbor did not think about tax incentives when she typed her question. She did not consider that Michigan's Senate Bill 237, signed eleven months earlier, helped determine where her tokens would be processed. She did not know that state legislators in Lansing had debated foregone revenue and job thresholds, that governors in a dozen states had competed to attract the facility that would serve her, that the race to the bottom had shaped a policy terrain she never sees.

The invisibility is the point. Good infrastructure disappears. The student experiences only the interface: a text box, a blinking cursor, words appearing as if by magic. The policy choices that enabled that magic—the tax exemptions, the ratepayer protections waived or imposed, the community benefits negotiated or foregone—exist in a world she has no reason to enter.

But the choices are real. Senate Bill 237 cost Michigan hundreds of millions in foregone revenue. The incentive package attracted billions in investment. The calculation—whether foregone taxes were worth the economic activity they enabled—will be debated for years. *Frank* and his fellow board members made their own calculations, weighing community values against legal costs, local control against economic pressure. They lost. The project proceeded. The incentives worked as designed.

We have traced her token through technology and geography, through silicon and concrete and copper. We have followed the money from Wall Street to Washtenaw County. Now the frame widens further. The student's question, answered by chips in Michigan, connects to competitions that span continents. The infrastructure serving her is not just an economic asset. In Washington and Beijing, it is a strategic one.

The next chapter examines a dimension of data center development even less visible than state incentives: the geopolitics that make AI infrastructure a matter of national security. From Michigan township boards to Pentagon planning offices, the stakes extend far beyond local tax policy.

CHAPTER TEN

The Geopolitics

O N JANUARY 21, 2025, PRESIDENT DONALD TRUMP STOOD IN THE ROO-
SEVELT ROOM. Executives from America's most powerful technology companies flanked him. Behind them, a screen displayed two words:
"Stargate Project." Reporters leaned forward. Cabinet officials stood
shoulder to shoulder with Silicon Valley billionaires. The tableau was
unusual, and it promised something consequential.

"These world-leading technology giants are announcing the formation of Stargate," Trump declared. "A new American company that will
invest five hundred billion dollars, at least, in AI infrastructure in the
United States... creating over a hundred thousand American jobs almost
immediately." He called it "the largest AI infrastructure project by far in
history."[55]

The executives stepped forward to speak. Larry Ellison described data
centers already under construction in Abilene, Texas. "Each building's a
half a million square feet," he said. "There are ten buildings currently
being built, but that will expand to twenty"—with other locations to follow. Sam Altman called it "the most important project of this era," adding
that he was "thrilled we get to do this in the United States of America."
Masayoshi Son of SoftBank outlined the financing: "We would immediately start deploying a hundred billion dollars, with a goal of making five
hundred billion within next four years." He named SoftBank, OpenAI, Or-

acle, and MGX—the UAE sovereign wealth fund—as partners, with Nvidia providing chips and Microsoft offering continued support.[2,196]

The full scope would emerge in subsequent filings: ten gigawatts of computing power planned by 2029. Enough to train AI systems that do not yet exist. Systems that will transform every dimension of economic and military power.

The announcement combined elements that rarely appear together in American politics. A Republican president celebrated private investment in industrial capacity, sounding themes Democrats more often claim. Technology executives spoke of national security and strategic competition with China—language from Pentagon briefings, not product launches. Foreign sovereign wealth funds pledged billions to support American AI dominance. Any one of these would dominate a normal news cycle. Together, they signaled a turning point.

This chapter traces how artificial intelligence infrastructure became a matter of national security. We follow the thread from semiconductor fabrication in Taiwan to server farms in Texas, from export control negotiations in Brussels to township board meetings in rural Michigan. The geopolitics explains why a $7 billion project in Saline Township concerns presidents and generals—and why decisions about power, land, and incentives carry implications far beyond the communities where facilities rise.

10.1 THE NEW STRATEGIC RESOURCE

Throughout the twentieth century, great powers competed for strategic resources. Oil fueled mechanized warfare in World War II and jet aircraft in the Cold War. Uranium enabled nuclear weapons that reshaped international relations. Semiconductors powered the digital revolution, transforming warfare as profoundly as the internal combustion engine a century earlier.

Artificial intelligence is the strategic resource of the twenty-first century.

The insight is not new. Pentagon planners have discussed AI's military applications for decades: autonomous weapons, surveillance, logistics, cyber operations, strategic planning. The nation that leads in artificial intelligence gains advantages across every dimension of national power—economic, military, diplomatic, and cultural.

What changed in the early 2020s was the recognition that AI capability depends on infrastructure. The transformer architecture introduced in 2017 created a framework where capabilities emerged from scale itself.[197] Larger models trained on more data with more compute demonstrated abilities that smaller models lacked. The scaling laws held across orders of magnitude. The race for AI capability became a race for compute.

And compute means data centers.

Training a frontier AI model requires thousands of specialized chips running continuously for months. The training run for GPT-4, completed in late 2022 or early 2023, reportedly used approximately 25,000 A100 GPUs running for three to four months.[198] Only a handful of facilities worldwide can house, power, and cool these chips at scale. Most organizations cannot access what frontier training demands.

GPT-5, released in August 2025, required computing resources orders of magnitude larger than GPT-4. Training the next generation—GPT-6 and beyond—will demand even more. The Stargate initiative is deploying 400,000 GPUs at its first site in Abilene, Texas—sixteen times the GPT-4 training cluster.[199] Across all planned Stargate sites, the project will deploy four to five million GPUs drawing ten gigawatts of power. This infrastructure is being built now, rapidly, at a scale that has no precedent in the history of computing.

The country that builds first gains an advantage that may prove impossible to overcome. Advanced AI systems could deliver military advantages in autonomous weapons, intelligence analysis, cyber operations, and strategic planning. Economic productivity gains would compound over decades. Scientific breakthroughs in materials science, drug discovery, and energy systems could reshape entire industries. Fall behind in AI capability and you fall behind permanently. The gap widens faster than any investment can close it.

This is why the White House hosted a press conference for what might otherwise seem like a private investment announcement. Why Congress authorized billions in subsidies for AI-related industries. Why the Department of Defense tracks data center construction with the same attention it pays to foreign military deployments. And why decisions in county planning offices carry national security implications that most local officials never imagined.

10.2   THE CHINA CHALLENGE

The strategic competition shaping AI infrastructure is bilateral: the United States versus China. Other nations matter—allies, competitors, battlegrounds for influence—but this central rivalry defines the field.

China's ambitions in artificial intelligence are explicit, documented, and well-funded. In 2017, the State Council published the "New Generation Artificial Intelligence Development Plan," setting a goal of global AI leadership by 2030.[200] The plan allocated billions in government funding, directed state-owned enterprises to prioritize AI, established development zones in major cities, and created incentives for private investment.

The timeline was ambitious but specific. Match developed countries by 2020. Achieve major breakthroughs and become a global leader in some AI fields by 2025. Become the world's primary AI innovation center by 2030, establishing international standards and leading theoretical development.

Chinese companies responded with massive investment. Baidu, Alibaba, Tencent, and Huawei built AI research laboratories that attracted top talent from around the world. Chinese researchers began publishing in top AI conferences at rates rivaling or exceeding American institutions. Universities expanded AI programs, graduating tens of thousands of machine learning specialists annually.

DeepSeek, founded in 2023, developed large language models that some independent benchmarks rank competitively with OpenAI's offerings.[201] Baidu's ERNIE Bot serves hundreds of millions of users. ByteDance's AI systems power recommendation algorithms reaching billions through TikTok and Douyin. Chinese AI has achieved capabilities that outside observers often underestimate.

China's "East Data, West Compute" strategy, announced in 2022, coordinated national data center investment through 2025.[202] The plan designated eight regional computing hubs, connected them with high-speed fiber, and directed workloads to western provinces where land is cheap and renewable energy abundant. At national scale, the strategy mirrors infrastructure planning that American companies must pursue without government coordination.

The investment was massive. Chinese sources estimated total data center infrastructure spending through 2025 at approximately $440 billion—comparable to American hyperscaler commitments. China built AI infrastructure at a scale that matched or exceeded American efforts in many dimensions. And China builds fast. As NVIDIA CEO Jensen Huang observed in late 2025: "China can build a hospital in a weekend. We take three years to build a data center."[203] Centralized authority, expedited approvals, and state-directed construction create speed that American regulatory processes cannot match.

But China faces constraints that the United States does not.

The most significant is access to advanced semiconductors. The most powerful AI chips—NVIDIA's H100, H200, and Blackwell series; AMD's MI300 line; Google's TPU accelerators—are manufactured using extreme ultraviolet lithography. Only one company makes EUV machines: ASML of the Netherlands. ASML supplies fabrication facilities in Taiwan (TSMC), South Korea (Samsung), and the United States (Intel). It does not supply China.

Without EUV equipment, Chinese semiconductor manufacturers cannot produce chips at the advanced nodes—7 nanometers, 5 nanometers, 3 nanometers—that enable the most capable AI accelerators. SMIC, China's leading chipmaker, has pushed older deep ultraviolet lithography to its limits, reportedly producing some 7-nanometer chips through heroic engineering. But 5 nanometers and beyond may be impossible without EUV.

This dependency creates vulnerability. If China cannot manufacture advanced chips domestically, it must import them. Imports can be blocked.

10.3 EXPORT CONTROLS

The United States recognized China's semiconductor dependency and moved to exploit it.

In October 2022, the Commerce Department's Bureau of Industry and Security issued export control rules restricting sales of advanced AI chips and chip-making equipment to China.[47] The rules targeted chips exceeding specified performance thresholds, effectively banning exports of NVIDIA A100s, H100s, and subsequent generations. They also restricted EUV lithography equipment, blocking China's path to domestic manufacturing at advanced nodes.

The October 2022 controls represented the most significant technology restriction the United States had imposed since the Cold War. Earlier export controls targeted specific military technologies or specific end users. The 2022 rules targeted entire categories of civilian technology based on their potential military applications. Nothing like it had been attempted before.

Jake Sullivan, the National Security Advisor, explained the strategy in a September speech: "On export controls, we have to revisit the long-standing premise of maintaining 'relative' advantages over competitors in certain key technologies. We previously maintained a 'sliding scale' approach that said we need to stay only a couple of generations ahead. That is not the strategic environment we are in today. Given the foundational nature of certain technologies, such as advanced logic and memory chips, we must maintain as large a lead as possible."[204]

The phrase "as large a lead as possible" signaled a strategic shift. American policy had long accepted that technology would eventually diffuse to competitors, seeking only a temporary advantage. The new approach aimed to maximize the gap and maintain it permanently—to ensure that China could never catch up in the most advanced semiconductor technologies.

China protested through diplomatic channels, lobbied American allies to maintain exports, and sought technical workarounds. Huawei developed its Ascend 910B chip as an alternative to NVIDIA products, achieving performance roughly comparable to the A100—a generation behind but still capable of meaningful AI training. SMIC pushed older lithography to its limits. Chinese companies stockpiled advanced chips before restrictions took full effect.

The United States responded with tighter controls. In October 2023, one year after the initial rules, the Bureau of Industry and Security issued updated regulations closing loopholes and expanding restrictions.[205] Chips designed specifically to avoid the original performance thresholds now fell under the ban. Third-country transfers that might enable Chinese access faced stricter scrutiny. The net tightened.

The Biden administration also pursued export control harmonization with allies. Without cooperation from the Netherlands (ASML's home) and Japan (Tokyo Electron's etching and deposition equipment), China could circumvent American restrictions by purchasing directly from non-American suppliers.

In early 2023, after months of negotiations, the Netherlands and Japan agreed to restrict their own exports of advanced semiconductor manufacturing equipment to China. The restrictions were not identical to American rules; each country maintained some discretion over implementation. But they closed the most significant loopholes. A technological bloc surrounding American AI and semiconductor leadership began to cohere.

By late 2024 and into 2025, the effects became visible. Chinese AI companies reported training runs constrained by chip availability. DeepSeek and others optimized model architectures to reduce compute requirements, achieving impressive efficiency gains that drew international attention. But these optimizations operated under fundamental hardware constraints that American competitors did not face.

The AI infrastructure race cannot be separated from this context. Every data center built in the United States represents computing capacity that China cannot easily match. Every gigawatt committed to American AI training compounds the advantage that export controls created. Every month of delay in Chinese semiconductor manufacturing extends the window of American dominance.

The chip war transformed the strategic significance of data center construction. What might once have seemed like commercial real estate development now carries implications for the global balance of power.

10.3.1   Policy Reversal

The export control regime that seemed stable through 2024 began shifting in 2025. The Biden administration, in its final days, issued an "AI Diffusion Rule" on January 15, 2025, establishing global performance thresholds and a tiered country access system. Tier 1 countries—the United States and eighteen close allies—faced no restrictions. Tier 2 countries faced quantity limits. Tier 3, including China, Russia, Iran, and North Korea, faced outright bans on advanced GPU exports. The rule was scheduled to take effect on May 15, 2025.

It never did. The Trump administration, returning to office in late January, initiated rescission proceedings. By May 12, 2025, the Bureau of Industry and Security formally withdrew the rule before it could take effect. The reversal signaled a different approach: the new administration sought flexibility in export controls, not maximum restriction.

The policy shift extended beyond rescission. On December 8, 2025, the Commerce Department approved exports of H200 chips to China, subject to a 25 percent revenue share with the United States government. The approach was novel: rather than banning sales, the government would take a cut. NVIDIA could sell to Chinese customers. American taxpayers would benefit from Chinese AI development. The restriction became a tariff.

Blackwell-generation chips remained prohibited. The distinction reflected a gap the administration considered strategically acceptable: H200s are powerful but represent the previous generation. Blackwell and its successor, Rubin, would remain restricted. The United States would maintain a lead on the cutting edge while monetizing technology that Chinese developers might otherwise obtain through other channels.

The policy drew criticism from multiple directions. National security hawks argued that any chip sales to China accelerated the adversary's AI development. Industry advocates countered that revenue sharing captured value from transactions that would happen anyway, whether through smuggling networks or domestic Chinese alternatives. Neither side was entirely satisfied.

The smuggling argument had empirical backing. Federal prosecutors had spent the previous year dismantling a network that moved approximately $160 million in H100 and H200 GPUs to China between October 2024 and May 2025. The chips traveled through Singapore and Malaysia, their end users obscured through shell companies and false documentation. The network's shutdown, announced in early December 2025, demonstrated both the demand for restricted chips and the difficulty of enforcing export controls at scale.

The Trump administration's approach reflected a calculation: better to capture revenue from controlled sales than to create black markets that captured nothing. Critics called it capitulation. Supporters called it pragmatism. The debate would continue as the policy took effect in early 2026.

The Bureau of Industry and Security simultaneously tightened enforcement in other areas. New guidance declared that using Huawei Ascend chips anywhere in the world—not just in China—violated American export controls. Supply chain diversion guidance warned companies that their chips could not end up training Chinese AI models, regardless of initial sale location. The combination of relaxation and restriction suggested strategic recalibration rather than wholesale retreat.

Congress responded to the executive branch's policy shift. On January 21, 2026, the House Foreign Affairs Committee voted 42 to 2 to advance the AI Overwatch Act, the first congressional vote on any legislation limiting AI chip sales to China.[206] The bill, sponsored by Chairman Brian Mast with bipartisan support from Ranking Member Gregory Meeks, would require the administration to give lawmakers thirty days to review chip exports to adversary nations before they occur. The process mirrors congressional oversight of arms sales, established in 1976. The bill would also ban sales of chips more advanced than the H200 or AMD's MI325x to China for twenty-four months, while streamlining exports to allied nations. "This bill is very simple," Mast said. "It keeps America's advanced AI chips out of the hands of Chinese commie spies." The bill must still pass the full House and Senate, but the overwhelming committee vote signaled bipartisan appetite for congressional involvement in export control decisions that the executive branch had treated as its exclusive domain.

10.4 THE STARGATE ANNOUNCEMENT

The Stargate initiative must be understood against this geopolitical back-drop.

When President Trump stood in the Roosevelt Room on January 21, 2025, he was not simply celebrating a large private investment. He was declaring American intent to win the AI infrastructure race—to build the computing capacity that would ensure American dominance in artificial intelligence for decades.

The $500 billion commitment was designed to grab headlines, but the specific terms revealed strategic priorities. OpenAI would serve as operational lead, defining technical requirements and consuming the vast computing capacity the project would create. Oracle would operate the data centers, providing infrastructure expertise. SoftBank and MGX would provide capital—foreign money accelerating American AI development. NVIDIA would supply the chips that actually perform AI computations.

The arrangement concentrated American technology leadership—OpenAI's models, NVIDIA's chips, Oracle's infrastructure—while drawing on foreign capital to accelerate deployment. Japanese and Emirati money would fund American AI advantage. The partners might profit handsomely, but the technological capability would remain in the United States.

Speed was the point. Stargate's first campus in Abilene, Texas, began construction in mid-2024, months before the formal White House announcement. It achieved initial operational status by September 2025.[199] Phase one deployed over 200 megawatts and installed GPU clusters capable of meaningful AI training. Phase two, targeting completion in 2026, would expand to 1.2 gigawatts and support 100,000 GPUs on a single high-speed network fabric—the largest coherent AI training cluster ever built.

The timeline was aggressive because the strategic logic demanded urgency. China was not standing still. Chinese data center construction continued despite semiconductor constraints. Chinese AI researchers

273

continued optimizing architectures for available hardware. Every month of delay was a month in which the American advantage might erode.

The national security connection was explicit. Sam Altman called Stargate "the most important project of this era" and emphasized that it would enable "AGI to get built here" in the United States—systems with obvious applications for intelligence analysis, strategic planning, and autonomous operations.[55] The White House fact sheet noted the project's importance for "maintaining America's competitive edge in artificial intelligence."

The Biden administration had made similar arguments about AI's strategic importance. The Trump administration's willingness to host the announcement reflected bipartisan recognition that AI infrastructure represented a national security priority transcending normal political divisions.

Saline Township entered this picture as a second-phase Stargate site. The Related Digital project represented capacity for OpenAI's next generation of AI training. Michigan's emergence as a data center destination followed directly from Stargate's infrastructure strategy: sites across multiple regions with available power, suitable land, and cooperative state governments.

Local residents learned, perhaps for the first time, that their township's zoning decisions connected to great power competition. The project they had voted to reject was, in Washington's view, a national strategic asset—one the federal government celebrated even as local residents protested.

One year after the initial announcement, OpenAI published community commitments responding to the concerns that had surfaced in Saline and elsewhere.[207] The company claimed to be "well beyond halfway" to its 10-gigawatt goal in planned capacity, with sites operating or under development in Texas, New Mexico, Wisconsin, and Michigan. The announcement pledged that Stargate would "pay its own way on energy" so operations would not increase electricity prices for existing customers. It promised closed-loop cooling to minimize water use, workforce development through "OpenAI Academies," and site-specific infrastructure investments. In Michigan, the company said, battery storage would be "financed entirely by the project" to ensure "no impact on DTE's existing customers' energy supply or rates."

Whether these commitments would survive contact with operational reality remained to be seen. OpenAI's track record on honoring public commitments offered grounds for skepticism. The company was founded in 2015 as a nonprofit dedicated to developing AI "for the benefit of humanity as a whole" and sharing research openly. By 2019 it had restructured as a capped-profit company. By 2023 it had stopped publishing research on its most capable models. By 2024 it was exploring conversion to a fully for-profit structure. Each transition was explained as necessary for the mission, but the pattern suggested that statements made in one context might not bind decisions in another. Elon Musk, an early funder who donated $44 million to OpenAI before departing the board in 2018, filed suit alleging the company had betrayed its founding promises. In January 2026, a federal judge denied OpenAI's motion for summary judgment, finding "ample evidence" that the company's leadership had said one thing publicly while planning something different privately. The evidence carried enough weight that Judge Yvonne Gonzalez Rogers sent the case to a jury convening in April 2026.[208]

The irony runs deeper. OpenAI's own inference demands—hundreds of millions of users, billions of daily queries—create the gravitational pull driving the infrastructure buildout that communities like Saline Township must accommodate. The same investors and stakeholders pressuring OpenAI toward profitability also expect returns from Stargate's massive capital deployment. Community commitments compete with quarterly earnings expectations. The forces that transformed OpenAI from nonprofit to profit-seeking enterprise are the same forces that will test whether Stargate's community pledges hold.

The Stargate pledges addressed legitimate concerns—but they came after approvals, not before. Communities that had accepted projects based on earlier, vaguer assurances had no mechanism to enforce new promises. The pattern of announcement, opposition, settlement, and then post-hoc community engagement left open questions about what bargaining power communities would retain if operators failed to deliver.

10.5 ALLIED COORDINATION

The United States does not compete with China alone. A network of allies shares American concerns about Chinese technological advancement and coordinates policies to address them. This coordination shapes where AI infrastructure gets built—and who can access it.

The EUV machines described in Chapter 2 represent extraordinary global manufacturing cooperation: Dutch optics, German precision engineering, Japanese chemical expertise, American software and design tools all contribute. No single country possesses all the knowledge and components to build them. In early 2023, the Netherlands and Japan agreed to restrict their own exports of advanced semiconductor manufacturing equipment, effectively embargoing the machinery required to manufacture cutting-edge AI chips.

Europe pursues "strategic autonomy," reducing dependence on American technology while maintaining the transatlantic alliance. But Europe cannot match American or Chinese scale. Of the projected $2 trillion in global data center investment through 2030, only about 24 percent flows to Europe. High energy costs, regulatory complexity across 27 member states, limited land, and political fragmentation constrain coordinated investment. Europe focuses on regulated workloads where data sovereignty justifies premium pricing, while the massive AI training infrastructure shaping frontier models rises primarily in the United States.

Japan, South Korea, Taiwan, and Australia align more closely with American strategy. These nations share concerns about Chinese technological and military advancement—Taiwan existentially so. They participate in technology cooperation frameworks and host American hyperscaler investment under security arrangements both sides accept.

The result is something approaching a technological bloc. Advanced semiconductors, manufacturing equipment, and AI chips concentrate in countries that share intelligence through Five Eyes—the alliance of the United States, United Kingdom, Canada, Australia, and New Zealand—or maintain close security partnerships with the United States. The geography of AI capability increasingly reflects political alignment rather than economic efficiency.

10.6 ENERGY SECURITY

Two additional dimensions matter: energy security and data sovereignty. Both shape where infrastructure gets built.

Energy security concerns arise from data centers' massive power consumption. A gigawatt-scale campus consumes as much electricity as a medium-sized city, every hour of every day, year after year. Countries hosting AI infrastructure must generate or import that power reliably. Energy vulnerabilities become AI vulnerabilities.

Europe experienced this acutely after Russia's invasion of Ukraine in February 2022. The cutoff of Russian natural gas sent electricity costs in Germany, France, and other major economies doubling or tripling. Data center operators watched their largest operating expense spike unpredictably. Some projects were delayed; others cancelled entirely. The crisis revealed how energy dependence constrains digital infrastructure.

The United States enjoys relative energy security that provides competitive advantage. Domestic natural gas production exceeds consumption, making America a net energy exporter. Nuclear plants, though aging, provide reliable baseload generation. Renewable capacity grows rapidly. Diverse supply sources and domestic production insulate American data centers from the price volatility that constrains European competitors.

Texas exemplifies the energy advantage. ERCOT's deregulated market and abundant natural gas enable electricity prices substantially below European or East Asian levels. Operators can contract directly with wind farms or gas generators, locking in long-term prices. The reliability concerns exposed by Winter Storm Uri are real but have not deterred operators who manage risk through backup generation.

China's energy position is mixed. The country generates more renewable electricity than any other nation and leads in nuclear plant construction. But China relies heavily on coal—approximately 55 percent of generation—and imports significant quantities of oil and natural gas. The energy mix constrains data center development in some regions while enabling it in others.

Data sovereignty concerns cut differently, creating both constraints and opportunities.

The European Union's General Data Protection Regulation (GDPR), effective since 2018, established the world's most stringent data protection regime. European citizens' data must remain in Europe or transfer only to countries with "adequate" protection—a status the United States lost in the Schrems II ruling of July 2020, when the European Court of Justice invalidated the Privacy Shield framework.

The ruling created legal complexity for American technology companies serving European customers. Some workloads became impossible to serve from American data centers. The practical result: pressure for data localization, keeping European data in European facilities regardless of where computation might most efficiently occur.

American hyperscalers responded by building substantial European operations. AWS, Microsoft Azure, and Google Cloud all maintain data center regions across Europe, enabling customers to keep data within EU boundaries. But the requirement fragments infrastructure, reducing efficiency and increasing costs. Storage for European users duplicates storage for American users. Processing that might be centralized must instead be distributed. The regulatory requirement imposes costs that all parties bear.

China takes data sovereignty further. The Cybersecurity Law of 2017, the Data Security Law of 2021, and the Personal Information Protection Law of 2021 collectively require that sensitive data remain within China's borders. Foreign companies must operate through local partners who ensure government access when requested. The Great Firewall filters content flowing into and out of China, restricting both information and data flows.

These regimes fragment the global internet into distinct zones with different rules and different relationships between companies and governments. AI models trained in one zone may not deploy in another. Data that could improve models cannot cross borders freely. The fragmentation imposes costs—industry estimates suggest 20 to 40 percent above unified global infrastructure—but reflects political choices that governments have made deliberately.

10.7 THE GLOBAL RACE

Beyond the US-China rivalry, a broader competition for AI infrastructure is reshaping the global technological order. The geography of that buildout reveals political alignment as much as economic logic.

The Middle East has emerged as a significant player, deploying sovereign wealth to position for a post-oil future. The UAE's MGX fund joined Stargate as a founding partner, committing $7 billion. Saudi Arabia's Public Investment Fund committed a reported $40 billion to AI-focused investment.[209] Oil-dependent economies recognize that AI investment hedges against energy transition risk while buying influence in an industry likely to reshape global power.

India has accelerated dramatically, emerging as a market neither superpower dominates. The government's Data Center Policy of 2020 granted infrastructure status to data centers, enabling tax benefits and expedited approvals. AWS, Microsoft, and Google have announced over $20 billion in collective Indian investment. India could host 10 to 12 percent of global data center capacity by 2030, with advantages including English proficiency, cost structures 30 to 50 percent below American levels, and a democratic government aligned with neither Beijing nor Washington.

Singapore illustrates the constraints success creates. By 2019, data centers consumed approximately 7 percent of national electricity, an unsustainable share for a 728-square-kilometer island. The government imposed a moratorium on new construction from 2019 to 2022.[210] Post-moratorium approvals require stringent efficiency standards; lower-margin compute increasingly moves to Malaysia or Indonesia.

The pattern holds across regions. Countries aligned with the United States host American hyperscaler investment under acceptable security arrangements. China builds within its own borders and in countries willing to accept Chinese terms. Non-aligned nations navigate between blocs, seeking investment from multiple sources while managing political complexities.

10.8 MILITARY APPLICATIONS

The connection between AI infrastructure and national security extends beyond economic competition to direct military and intelligence applications.

Autonomous weapons represent the most visible application. Drones that select and engage targets without human intervention depend on AI for image recognition, path planning, and decision-making. Intelligence analysis increasingly relies on machine learning to process volumes of information no human analyst could review manually: satellite imagery interpretation, signals intelligence, open-source aggregation across social media and financial data. Logistics, cyber operations, and strategic planning all benefit from AI capabilities that scale with compute.

The more capable the AI, the more effective these applications. Systems trained on massive datasets using infrastructure concentrated in a handful of data centers gain advantages that less well-resourced competitors cannot match.

The United States maintains classification regimes that formally separate military and civilian AI development. But the underlying technologies overlap substantially. A language model trained for commercial applications could support intelligence analysis. An image recognition system for autonomous vehicles could identify military equipment from satellite imagery. The "dual-use" nature of AI makes clean separation impossible.

This creates policy complications. The infrastructure that trains ChatGPT could train systems with national security purposes. Stargate, though framed as commercial, enables capabilities the Department of Defense values and monitors closely. Technology companies emphasize civilian use cases in public communications, but potential military utility shapes how national security establishments perceive AI infrastructure— regardless of what developers intend.

The separation grew more porous through 2025. In July, the Pentagon's Chief Digital and Artificial Intelligence Office awarded contracts worth up to $200 million each to OpenAI, Anthropic, Google, and xAI— frontier AI companies that had previously emphasized civilian applications.[211] The contracts support development of AI agents for military planning and operations. OpenAI, which had explicitly prohibited military use of its technology until January 2024, now builds tools for the Department of Defense.

In October, the Air Force invited private companies to build AI data centers on military bases, offering fifty-year leases across 3,100 acres at five installations.[212] The requirements—minimum 100 megawatts, $500 million investment—suggest facilities comparable to commercial hyper-scale campuses. The government retains options to purchase services from these privately operated facilities. The line between commercial infrastructure and military asset blurs further with each announcement.

Strategic planners in Washington, Beijing, and Moscow factor AI infrastructure into assessments of relative national capability. The data centers rising across America are, from their perspective, assets enhancing national power. What operators intend matters less than what the infrastructure enables.

10.9 THE REGULATORY DIMENSION

Governments worldwide are regulating AI development in ways that affect infrastructure requirements. The regulatory environment evolves rapidly.

The European Union's AI Act, entering force progressively through 2026, classifies AI systems by risk level and imposes corresponding requirements.[213] High-risk systems—those used in critical infrastructure, law enforcement, employment, or education—must meet standards for transparency, accuracy, human oversight, and documentation. General-purpose AI models with systemic risks face additional requirements, including disclosure of training methodologies and energy consumption.

The AI Act's infrastructure implications are significant. Compliance requires retaining training data, model versions, and audit logs potentially for years. Industry estimates suggest storage needs rise 20 to 30 percent above baseline for affected workloads. Data centers must accommodate the additional capacity.

The United States took a lighter approach. President Biden's Executive Order 14110, issued in October 2023, required reporting of large training runs but stopped short of mandatory standards.[214] The Trump administration, returning to office in January 2025, emphasized facilitating AI development rather than constraining it.

The regulatory divergence creates geographic incentives. Companies seeking to avoid European compliance costs may locate AI training in the United States. Companies serving European customers may find European data centers advantageous for demonstrating compliance with AI Act requirements applying extraterritorially.

China's AI regulations embed Communist Party priorities. Content generated by AI must align with "socialist core values." Large models require registration with the Cyberspace Administration before public deployment. Training data and model weights may be subject to government access.

The Chinese approach creates an AI environment distinct from the rest of the world, optimized for state oversight rather than open innovation. Chinese companies must navigate political requirements that American and European competitors do not face. The burden creates friction that may slow development—but also creates a market foreign competitors cannot easily serve.

The fragmentation of AI regulation mirrors the fragmentation of data protection regimes. Companies operating globally must navigate multiple frameworks with different requirements and enforcement mechanisms. The compliance burden falls particularly hard on smaller players lacking resources for multi-jurisdictional legal expertise.

10.10 WORKFORCE AND HUMAN CAPITAL

AI infrastructure development faces workforce constraints that cross national boundaries and shape competitive dynamics—often overlooked in strategic discussions focused on hardware and capital.

Building and operating data centers at frontier scale demands specialized skills in short supply globally. Power systems engineers who can manage gigawatt-scale electrical infrastructure. Cooling specialists who understand heat removal from densely packed GPU clusters. Fiber network experts who can provision high-bandwidth, low-latency connectivity. Security professionals who can protect facilities housing billions in equipment.

The talent pool is limited globally and effectively fixed in the short term. The United States trains many engineers, but graduates compete with technology companies, utilities, and defense contractors for their services. Wages for data center professionals have climbed sharply as demand outpaces supply. Even at elevated wages, positions remain unfilled for months.

Immigration policy intersects with workforce development in ways that shape American competitive position. Many engineers working in American data centers were born abroad and entered through employment-based visa programs. The H-1B program, despite controversies, supplies significant numbers of technology workers who cannot easily be replaced domestically.

Restrictions on immigration have reduced the available workforce for AI infrastructure development. The constraint is not hypothetical. Industry surveys consistently rank workforce availability among the top constraints on data center expansion, alongside power availability and permitting timelines. Tighter visa programs would exacerbate an already acute shortage.

China graduates enormous numbers of engineers—more than the United States by most measures. But skills for hyperscale infrastructure operation do not necessarily match what Chinese universities teach. Practical experience at frontier scale exists primarily in American companies operating the world's largest data centers. Chinese operators are developing expertise, but the learning curve is steep.

India similarly graduates large numbers of technical workers. But operating complex infrastructure requires more than degrees. It requires institutional knowledge about what can go wrong and how to prevent failures—knowledge developed through years of operational experience.

Geographic concentration of AI expertise creates regional advantages. Northern Virginia's deep talent pool, developed over three decades of data center operation, attracts continued investment because qualified workers live nearby. New regions entering the market—including Michigan, Kansas, and other states pursuing aggressive incentives—must develop workforce capacity alongside physical infrastructure. That process takes years.

Training programs are emerging to address the shortage. Community colleges offer data center technician certifications. Hyperscalers run their own training academies. Union apprenticeship programs are adapting to include data center skills.

But workforce development takes time that the AI infrastructure race may not allow. Facilities under construction will need trained operators before training programs graduate their first cohorts. Capital alone cannot solve this bottleneck.

10.11   THE SALINE TOWNSHIP CONNECTION

The geopolitics traced in this chapter connect directly to what happened in Saline Township.

The Stargate initiative created demand for data center sites beyond Texas. OpenAI needed computing capacity across multiple geographic locations. Oracle and SoftBank had committed capital that required deployment. The partnership sought sites with available power, suitable land, cooperative state governments, and grid infrastructure capable of supporting gigawatt-scale loads.

Michigan offered what the partnership needed. DTE Energy could provide 1.4 gigawatts, drawing on available MISO capacity. Agricultural land outside Ann Arbor was available at prices far below Virginia or Texas markets. Senate Bill 237, signed in December 2024, demonstrated state support.[80]

When Governor Whitmer announced the project in October 2025, she emphasized its national significance.[18] The investment was not merely an economic development win for Michigan. It was a contribution to American AI leadership in competition with China.

Local opposition was overridden in part because of this context. A township board's rejection of a project that federal officials considered strategically important created tension that existing law could not easily resolve. The legal system resolved the conflict in the developer's favor,[56] but the underlying dynamic reflected a collision between local governance and national strategic priorities.

The community's concerns about electricity consumption, environmental impact, water usage, traffic, noise, and rural character remained valid on their own terms. But they competed with considerations extending far beyond Washtenaw County. Saline Township's land use decisions connected to great power competition, export controls, and military applications that local officials had neither the authority nor the information to evaluate.

This is the new reality of AI infrastructure. Decisions that once seemed purely local now carry implications that policymakers in Washington, Beijing, and Brussels track closely. The infrastructure enabling American AI capability must be built somewhere. Communities that host it participate—willingly or not—in the competition shaping the twenty-first century.

10.12   Future Trajectories

Where does the geopolitical competition for AI infrastructure lead?

Several trajectories seem plausible, though prediction is hazardous in a rapidly evolving field.

The US-China rivalry continues, though in forms that 2024 observers might not have anticipated. The December 2025 H200 approval suggests that maximum restriction is not the permanent posture. Export controls may evolve through cycles of tightening and relaxation as administrations change and circumstances shift. The revenue-sharing model could expand to other chip categories—or collapse if Chinese alternatives render American chips unnecessary. Technology decoupling remains possible but is no longer the default trajectory. The prospect of conflict over Taiwan, where TSMC's fabrication facilities sit, adds risk that strategic planners cannot ignore.

Allied coordination will probably deepen. Countries aligned with the United States share interests in maintaining advantage over China. Export control harmonization, intelligence sharing, and joint development programs will likely expand. The technological bloc could solidify into something resembling an AI alliance structure.

Energy constraints will shape infrastructure development for decades. Countries with abundant, reliable electricity will attract investment; those without will fall behind. Renewable intermittency challenges 24/7 power requirements, but nuclear and geothermal offer baseload alternatives that some regions are pursuing aggressively.

Regulatory divergence may accelerate or moderate depending on experience. If the EU AI Act proves unworkable, it could be revised. If American development produces harms that European regulations would have prevented, pressure for American regulation could mount.

New entrants could disrupt established patterns. India's rapid growth could position it as a hub serving neither superpower exclusively. Middle Eastern sovereign wealth could finance development in unexpected locations. Breakthrough technologies could shift competitive dynamics in ways favoring different geographic distributions.

Throughout these potential futures, one connection will remain: AI infrastructure is national power. The countries that build data centers will train AI systems. Those systems will shape economies and militaries. The race continues. The stakes justify the attention of presidents and generals—and of township board members who find themselves, unexpectedly, participants in great power competition.

10.13 IMPLICATIONS

This chapter has traced the geopolitics of AI infrastructure from the White House to Saline Township. Several implications emerge.

First, data center development is now a matter of national security. Federal officials track major projects, encourage investment through policy signals and subsidies, and may intervene when local decisions conflict with strategic priorities.

Second, the US-China rivalry shapes the industry's direction. Export controls, technology decoupling, allied coordination, and infrastructure racing all flow from great power competition. Companies and communities operating in this space must understand the broader context.

Third, allied coordination matters more than casual observation suggests. American AI leadership depends on partnerships with countries supplying semiconductor manufacturing and participating in export control regimes. Maintaining these relationships is a national priority.

Fourth, regulation fragments the global AI industry into distinct zones. European, American, and Chinese approaches differ substantially, increasing costs while creating opportunities for regulatory arbitrage.

Fifth, energy and data sovereignty create both vulnerabilities and opportunities. Countries with energy security and favorable data protection frameworks attract investment. Those without face disadvantages that capital alone cannot overcome.

Sixth, workforce constraints cross national boundaries regardless of capital availability. The limited talent pool for building and operating hyperscale infrastructure constrains all players.

These implications suggest that questions raised in earlier chapters—about power, land, money, and incentives—cannot be answered without reference to geopolitics. The infrastructure making artificial intelligence possible is strategic infrastructure. Its development will shape the balance of power for decades to come.

The next chapter turns from geopolitics to technology. What designs are emerging for the next generation of data centers? New cooling approaches, new power configurations, new architectures—all may change how AI infrastructure is built.

———

The student in Ann Arbor closed the laptop hours ago. The coffee shop is dark now, chairs stacked on tables, the staff long gone. But the infrastructure that answered that evening's questions never sleeps. In Saline Township, the servers run through the night, processing queries from users across time zones, training models that will be slightly better tomorrow than they were today.

The token's journey traversed more than physical infrastructure. It traveled through a web of international relationships that determines where AI capability can exist. The student's question, asked in Michigan, was answered by chips designed in California, manufactured in Taiwan, installed in servers assembled in Texas, deployed in a facility built by New York developers, financed partly by Emirati sovereign wealth, operating under American export controls designed to maintain advantage over China.

Each layer of that stack—silicon, power, land, capital, policy—now connects to questions of national security and great power competition.

This is what it takes to make words appear on a screen. This is the infrastructure beneath the interface. Trillions of dollars, hundreds of gigawatts, thousands of acres, millions of workers, and the strategic calculations of great powers—all converging so that a student in a coffee shop can ask a question and receive an answer that feels, impossibly, like magic.

The book's remaining chapters look forward. What new technologies might change this picture? What scenarios could unfold? And what does it all mean for the communities, the workers, and the nation that hosts this unprecedented buildout?

The global distribution of AI infrastructure capacity reflects this competition. The United States accounts for approximately 38 percent of global data center capacity, down from 42 percent in 2020 as other regions expand. Europe holds steady at about 24 percent. Asia-Pacific has grown from 28 percent to 32 percent, with India's capacity expanding 87 percent between 2022 and 2025. These shifts will accelerate as the projected $2 trillion in global investment through 2030 flows disproportionately toward regions with favorable power, policy, and workforce conditions.

CHAPTER ELEVEN

Future Designs

THE COOLING SYSTEMS DESCRIBED in Chapter 3 handle the rack densities that current AI accelerators demand. The next generation of chips will push beyond these limits. What comes next? Immersion systems capable of 300 kilowatts or more per rack. Modular construction that cuts build times from years to months. And the broader question of whether innovation can solve the problems that technology has created.

11.1 BEYOND CURRENT LIMITS

Even direct-to-chip cooling has limits. These systems still leave other components—memory, storage, networking equipment—cooled by air. Immersion takes a more radical approach: submerge the entire server in dielectric fluid that does not conduct electricity. Two-phase immersion systems use fluids engineered to boil at precisely controlled temperatures, absorbing enormous energy through phase change. A single liter of two-phase coolant absorbs over a hundred times more heat than a liter of air at the same temperature differential.

GRC, LiquidCool Solutions, Submer, and Asperitas have developed commercial immersion systems. The obstacles remain formidable: specialized server designs, complex maintenance, expensive fluids, uncertain long-term reliability. But several operators have announced immersion-cooled facilities for 2026-2028 deployment. If the technology proves reliable at scale, it could enable power densities exceeding 500 kilowatts per rack.

———

The calculation operators face: advanced cooling technologies exist, but deploying them at scale introduces risk. Data center operators are conservative by nature; their customers demand uptime measured in fractions of a percent. Facilities designed today must run reliably for fifteen to twenty years. Betting on unproven technology can cost more than accepting lower efficiency from proven approaches.

This conservatism collides with the accelerating pace of AI development. The chips being installed today will likely be superseded within two years. The cooling infrastructure must handle not only current equipment but whatever comes next. Designing for an uncertain future requires engineering margins that add cost without delivering immediate benefit—but avoids far more expensive reconstruction later.

11.2   MODULAR CONSTRUCTION

Time is money in data center development, but the calculation runs more extreme than in other industries. Every month of delay represents lost revenue that can never be recovered. An AI training cluster that comes online six months late misses a generation of model development. Computational advantage translates directly into commercial success. Speed matters as much as efficiency.

Traditional data center construction takes two to four years from first shovel to operation.[215] Site preparation, foundation work, building construction, electrical infrastructure, mechanical systems, security installations, equipment commissioning—each requires sequential completion. Delays compound. A problem in foundation work pushes back everything that follows.

Modular construction offers an alternative. Rather than building on-site from scratch, factories fabricate major components in controlled environments. These modules—complete with electrical connections, cooling systems, and equipment mounting—ship to the site for assembly. Construction becomes assembly.

The benefits extend beyond speed. Factory fabrication allows tighter quality control than field construction. Modules can be tested before shipping, catching problems before they create site delays. Labor productivity runs higher in factories than on construction sites. Weather delays vanish when most work happens indoors.

Steve has watched this transformation reshape his work over the past decade. "When I started in data centers, we built everything on-site," he says. "Poured the foundations, set the steel, ran the conduit—real construction. Now half my job is supervising crane operators unloading modules that somebody else built in a factory."

He does not resist the change. Thirty-two years in the trades have taught him to adapt. "The work follows the technology," he observes. "Used to be I needed crews who could do everything. Now I need crews who can connect things that come pre-made. Different skills. Faster timelines. Less margin for error because everything has to fit together precisely."

The implications for workforce development run deep. The construction trades that built traditional data centers—electricians, pipefitters, ironworkers—remain essential but in different proportions. New specialties emerge: logistics coordinators, integration technicians, quality inspectors focused on modular interfaces. The labor market for data center construction evolves as rapidly as the technology it serves.

Several companies have industrialized modular data center production. EdgeCore, backed by Mount Elbert Capital Partners, built a manufacturing facility in Nevada that produces standardized data hall modules.[216] Each module contains complete power distribution, cooling, and server mounting systems. Modules stack together like building blocks, creating facilities of varying sizes.

Compass Datacenters, headquartered in Dallas, operates what it calls a "factory for data centers" in Red Oak, Texas.[217] The company fabricates electrical and mechanical systems in standardized configurations that ship to project sites. This approach cuts typical construction schedules by forty percent.

Blackstone's QTS developed a modular platform called QTS Hyperscale that deploys preconfigured computing environments.[152] The company brings new capacity online in months rather than years. This speed has made QTS a preferred partner for hyperscale customers with urgent capacity needs.

The modular approach fits the current AI boom particularly well. Demand is growing faster than anyone predicted, and customers cannot wait years for capacity. Operators that deliver megawatts of AI-ready infrastructure in twelve months rather than thirty-six capture market share from slower competitors.

Modular construction cannot solve every challenge. The utility interconnection timelines described in Chapter 4 often exceed construction schedules. A modular building means nothing without electricity to power it.

Related Digital's construction plan includes significant modular components, allowing data halls to rise faster than conventional methods would permit. But the overall timeline depends on DTE Energy's infrastructure upgrades, which follow utility schedules rather than construction schedules.

Modular construction also assumes standardized requirements. It works well when customers want essentially identical capacity—which describes most hyperscale cloud deployments. It works less well when customers need customized configurations for specialized workloads. AI infrastructure falls somewhere between: standardized enough to benefit from modular approaches, specialized enough to require significant customization.

The industry is developing hybrid approaches that combine modular efficiency with customization flexibility. Factory-built components handle standard elements—power distribution, cooling loops, structural supports—while site-specific work addresses specialized requirements. This "mass customization" model borrows from automotive manufacturing, where standardized platforms support differentiated products.

11.3   EDGE VERSUS CENTER

Not all AI computation happens in hyperscale data centers. A growing category of workloads runs at the edge. These are smaller facilities distributed across metropolitan areas, embedded in commercial buildings, integrated with telecommunications infrastructure. The tension between centralized hyperscale and distributed edge shapes the industry's architectural future.

———

Training workloads strongly favor centralization. Physical proximity minimizes communication latency. A cluster spread across multiple facilities would spend more time waiting for data than computing with it. The power economics and land availability explored in earlier chapters reinforce concentration in rural locations.

———

Inference has different requirements. These workloads are latency-sensitive; users notice delays measured in milliseconds. Every mile between user and computation adds delay. For real-time applications like voice assistants or autonomous vehicles, delay is not merely inconvenient but functionally limiting.

Edge computing places inference capacity closer to users. A small facility in a metropolitan area serves local users faster than a distant hyperscale campus. A network of edge facilities provides geographic redundancy; if one site fails, nearby sites absorb its traffic. For applications where latency or reliability matters most, edge deployment offers advantages that centralized facilities cannot match.

The edge opportunity has attracted major investment. Vapor IO operates a network of small data centers deployed at cellular tower sites.[218] EdgeConneX develops mid-sized facilities in metropolitan markets. Amazon Web Services' Local Zones and Microsoft Azure's Edge Zones place cloud computing closer to end users. These investments bet that edge will grow alongside, not instead of, centralized hyperscale.

The emerging architecture combines hyperscale facilities for training and batch processing with distributed edge facilities for latency-sensitive inference. This hybrid model requires coordination between facilities at vastly different scales. Models trained centrally deploy to edge locations. Data flows between edge and center for analytics and model updates.

The technical challenges of hybrid deployment are substantial but solvable. More interesting are the economic implications. A hybrid architecture suggests that AI infrastructure will not concentrate indefinitely in a handful of gigawatt campuses. The industry may instead evolve toward a tiered structure: massive training facilities in locations like Saline Township, surrounded by networks of smaller inference facilities spread across population centers.

This evolution would change the community impact profile of AI infrastructure. Rather than a few communities hosting billion-dollar campuses, many communities might host smaller facilities. The power requirements of individual edge sites—typically measured in megawatts rather than gigawatts—fit more easily into existing grid infrastructure. The visual and environmental footprint shrinks proportionally.

But edge deployment complements hyperscale facilities rather than replacing them. The training complexes in rural Michigan or Kansas or New Mexico will continue operating and expanding. The question is whether future growth concentrates further in these locations or distributes across additional tiers of infrastructure.

A third tier has emerged between hyperscale and edge: the behind-the-meter hypergrid. These facilities—Oracle's VoltaGrid partnership, xAI's Colossus, the Joule project in Utah—generate their own power at gigawatt scale, operating semi-independently of public utilities. They represent neither centralized grid-connected infrastructure nor distributed edge computing, but a new category: privately powered AI factories that happen to connect to the grid for backup rather than primary supply. This architectural pattern may prove transitional, a bridge until grid capacity catches up with demand. Or it may become permanent, with the largest AI operators effectively becoming their own utilities.

————

Edge data centers represent one tier of distribution. A more radical shift is already underway: inference running directly on phones, laptops, and enterprise servers. The device in your pocket is becoming a data center.

Computing has seen this pattern before. The room-sized mainframes of the 1960s delivered fewer floating-point operations per second than a modern smartphone. What once required a climate-controlled facility with dedicated staff now fits in your pocket, powered by a battery.

The trajectory from mainframe to minicomputer to personal computer to smartphone traced a consistent arc: centralized resources became distributed capabilities as hardware improved.

The same cycle has repeated in software architecture. The original mainframe model was centralized: dumb terminals connected to shared computers. Personal computers shifted processing to the edge—thick clients running local software. Then the web pushed computation back to servers, with browsers as thin clients. Cloud computing extended this centralization. Now on-device AI may swing the pendulum again, returning intelligence to the edge. Each swing responds to the technology constraints of its moment. The current assumption that AI requires massive centralized infrastructure may prove as temporary as the assumption that computing required room-sized machines.

Apple Intelligence handles most requests on-device using a three-billion-parameter model; complex queries route to Apple's Private Cloud Compute.[219][220] Google's Gemini Nano runs on Android devices without network connectivity.[221] Qualcomm's Snapdragon X2 Elite delivers 80 trillion operations per second through its NPU, running models locally that would have required cloud infrastructure two years ago.[222]

Enterprise self-hosted inference has gone mainstream. vLLM provides production-grade serving used by IBM and Mistral AI.[223] The llm-d distributed stack enables Kubernetes-native deployment at enterprise scale.[224] The enterprise LLM market grows at 26 percent annually, projected to reach 71 billion dollars by 2034.[225]

What this means for hyperscale data centers depends on how much inference actually moves to devices and enterprise servers. Deloitte projects inference will represent two-thirds of all compute by 2026.[36] Training will remain centralized—large models require thousands of coordinated GPUs—but models are increasingly designed to run inference elsewhere. McKinsey projects inference growing at 35 percent annually to reach 93 gigawatts by 2030.[44] The trillion-dollar buildout continues.

But the architecture of AI computing may prove more distributed than current projections assume.

11.4 EXPERIMENTAL FRONTIERS

Beyond incremental improvements, researchers are exploring radically different approaches to data center architecture. Most will never achieve commercial viability. But understanding the frontier reveals the range of possibilities.

Microsoft's Project Natick deployed a sealed data center capsule on the seafloor off Scotland's Orkney Islands from 2018 to 2020. The stable temperature and absence of oxygen-driven corrosion yielded failure rates one-eighth those of comparable land-based equipment.[226] But manufacturing pressure vessels, deploying them to the seafloor, and servicing equipment remain prohibitively expensive. Microsoft has not announced plans for commercial deployment.

Space-based computing holds similar appeal—vacuum enables efficient heat radiation—and faces similar obstacles. Launch costs have fallen dramatically, with SpaceX's Falcon 9 delivering payloads for roughly two thousand dollars per kilogram.[227] Lumen Orbit, a startup founded in 2024, pursues orbital computing for specific AI workloads.[228] But latency, radiation damage, and the difficulty of equipment replacement make orbital data centers impractical for mainstream commercial use.

These experimental approaches illustrate a pattern: radical concepts offering genuine cooling advantages often cannot overcome economic barriers. The industry's future will more likely emerge from practical improvements to terrestrial facilities than from undersea or orbital deployments.

More practical than underwater or orbital facilities—though still at the frontier—are data centers with dedicated nuclear power generation. The nuclear partnerships described in Chapter 6—Microsoft's Three Mile Island restart, Amazon's X-energy agreement, Google's Kairos Power deal—reflect this path.[229][114][113]

The timeline challenges remain severe. Chapter 6 detailed the eight-to-twelve-year lead times for small modular reactors. In an industry where eighteen months feels like forever, decade-long timelines strain commercial planning. But if SMR technology matures as proponents hope, nuclear-powered data centers could become commonplace by the mid-2030s—addressing concerns about carbon emissions while introducing controversies of their own.

11.5 SUSTAINABILITY INNOVATION

The environmental impact of data centers drives urgent innovation. Operators face pressure from customers demanding carbon-neutral computing, regulators imposing emissions requirements, and communities questioning whether local resources should power AI systems. These pressures accelerate development of technologies that might otherwise have taken decades to mature.

———

Data centers convert electricity to heat with near-perfect efficiency. Every watt that enters eventually leaves as thermal energy. The question is whether that heat can do useful work instead of simply dissipating into the atmosphere.

Waste heat recovery captures thermal energy from data center cooling systems and applies it to productive purposes. In cold climates, recovered heat warms nearby buildings. In agricultural contexts, it heats greenhouses. Industrial processes requiring moderate temperatures can use data center waste heat as a free input. Chemical production, food processing, aquaculture: all are candidates.

The technical challenges are manageable. Data center cooling produces heat between ninety and one hundred forty degrees Fahrenheit. Hot enough for space heating and some industrial processes, though not for high-temperature applications. Heat exchangers transfer energy from cooling loops to district heating systems or industrial processes. The infrastructure for heat transport is well understood: insulated pipes carrying hot water.

The economic and logistical challenges prove harder. Waste heat has value only if someone nearby can use it. A data center in rural farmland has few heat customers. Even in urban areas, connecting data center cooling to heating systems requires coordination that rarely happens spontaneously. Building operators and data center operators answer to different priorities, often different owners.

Some European facilities have overcome these challenges. Stockholm Data Parks connects data centers to the city's district heating system, displacing fossil fuel heating for thousands of apartments.[230] Helsinki has similar arrangements. These northern European cities possess extensive district heating infrastructure, cold climates that create year-round heating demand, and regulatory environments that encourage the necessary coordination.

American data centers rarely achieve comparable integration. Most facilities simply exhaust heat to the atmosphere. The Saline Township project includes no waste heat recovery; no nearby facilities could use the thermal output economically. This represents a missed opportunity, but capturing it would require changes in land use planning, district heating infrastructure, and intercompany coordination that no single operator can achieve alone.

The Deep Green project in Lansing offers a different model. The data center solves the city's failing steam infrastructure problem; the city's heating demand makes the data center's waste heat valuable. Neither party could achieve this outcome alone. The model works because it matches facility scale to urban infrastructure needs—and because Lansing still has district heating infrastructure that most American cities abandoned decades ago.

One company is experimenting with a different approach. Bitzero, operating from a repurposed Cold War radar installation in North Dakota, uses server waste heat to warm an on-site greenhouse. The facility grows vegetables year-round, powered by excess thermal energy that would otherwise escape to the sky. This agricultural integration creates local value from what is typically an externality. Whether the model can scale to gigawatt facilities remains unclear.

Beyond energy efficiency, sustainability innovation increasingly focuses on material flows. Data centers consume vast quantities of steel, concrete, copper, and rare earth elements. The equipment they house—servers, storage devices, networking gear—cycles through rapid obsolescence. Sustainable operations require attention to these material impacts.

Server refresh cycles have shortened as AI workloads demand ever-more-capable hardware. Equipment that once operated for five to seven years may now be obsolete in two to three. Responsible operators must manage this hardware carefully: refurbishing and reselling equipment where markets exist, recycling materials where they do not, disposing of hazardous components properly.

Some companies are developing circular economy models specific to data centers. Iron Mountain, traditionally a document storage company, expanded into data center services with an emphasis on end-of-life hardware management. Sims Lifecycle Services specializes in processing retired equipment for material recovery. These businesses create markets that make responsible disposal economically viable.

Construction materials present similar challenges and opportunities. Data centers require vast amounts of concrete and steel, whose production generates substantial carbon emissions. Using recycled materials, specifying low-carbon concrete formulations, and designing for eventual deconstruction can reduce lifecycle emissions. Some operators now specify "embodied carbon" limits in construction contracts, pushing suppliers toward lower-impact materials.

The Saline Township facility incorporates some circular economy practices—Oracle requires documented electronics recycling for retired equipment—but falls short of the most ambitious sustainability goals. Comprehensive circular economy approaches require supply chain transformations that no single project can mandate.

Water consumption has become a flashpoint for data center opposition, particularly in water-stressed regions. Traditional evaporative cooling consumes millions of gallons annually. In drought-affected areas, this creates direct competition with agricultural, residential, and environmental water needs.

The industry responded with rapid adoption of waterless or reduced-water cooling technologies. These systems reject heat to the atmosphere through air-cooled heat exchangers rather than evaporation. The thermodynamic efficiency is lower—air carries heat less effectively than evaporating water—but the water savings are dramatic.

Edged Data Centers developed what it calls ThermalWorks, a proprietary waterless cooling system that the company claims saves over a hundred million gallons annually per facility.[231] CyrusOne operates waterless facilities in Arizona, where water scarcity makes traditional cooling politically untenable.[232] Aligned Data Centers' Delta3 technology reduces water consumption by eighty-five percent compared to conventional approaches.[233]

The trade-off is energy consumption. Waterless cooling systems require more fan power to move air across heat exchangers. In hot climates, energy consumption rises significantly during summer months. Operators must balance water savings against energy costs and carbon emissions from additional electricity consumption.

In moderate climates like Michigan's, the trade-off proves more favorable. Closed-loop systems minimize water loss while maintaining energy efficiency. The optimal cooling strategy varies by geography: what works in temperate Michigan differs from what makes sense in arid Arizona.

The water question illustrates how sustainability trade-offs resist simple solutions. Eliminate water consumption and you may increase carbon emissions. Minimize carbon and you may need nuclear power, with its own controversies. Reduce land use and you may increase community impact through denser development. These trade-offs demand choices. Different communities will reasonably make different ones.

11.6 WHAT SALINE'S FUTURE HOLDS

The technologies described in this chapter will reshape Saline Township's data center over its operational lifetime. The facility being built now represents 2025 design practice. By 2035, it will host equipment and employ techniques that do not yet exist. The building may look the same from the road; inside, everything will have changed.

Within five years, increased AI chip power consumption will likely drive deployment of advanced cold plates and possibly immersion systems for the highest-density racks.

Modular expansion is all but certain. Related Digital's development plans call for phased buildout that will add capacity over several years. Each phase will incorporate lessons from previous deployments and adapt to evolving requirements. The standardized infrastructure—power distribution, cooling backbone, network connectivity—provides a foundation for expansion that can proceed faster than original construction.

Energy mix evolution will reshape the facility's environmental profile. DTE Energy's service territory is adding renewable generation capacity, and the utility has committed to reducing carbon intensity over time. By 2030, a larger fraction of Saline's electricity will flow from solar, wind, and nuclear sources. Oracle's sustainability commitments will push for this transition; the company pledged to achieve net-zero emissions across its operations.[234]

Looking to 2035, more speculative changes become plausible. If small modular reactors achieve commercial deployment, Saline Township could see proposals for nuclear generation in the region. Michigan's policy framework, shaped by the state's nuclear history and the economic pressures of AI infrastructure, might accommodate such development.

Edge infrastructure could proliferate around the Saline campus. The hyperscale facility would serve as a hub, with smaller inference-optimized deployments distributed across southeastern Michigan. This tiered architecture would keep latency-sensitive workloads close to users while centralizing training and batch processing.

Waste heat recovery might finally become practical as the surrounding area develops. If the region's population grows—attracted by jobs at the data center and related industries—district heating systems could emerge that capture thermal energy currently exhausted to atmosphere. This evolution would require planning and coordination that has not yet begun. The opportunity sits idle, waiting for someone to seize it.

By 2040 or 2050, the Saline Township facility might be unrecognizable compared to its current design. The building shells may persist, but the equipment, cooling systems, and operational practices will have cycled through multiple generations. Technologies that seem exotic today—immersion cooling, onsite nuclear, comprehensive heat recovery—may be standard practice.

Or the facility might be obsolete. Technological disruption could render current architectures uncompetitive. Quantum computing, neuromorphic processors, efficiency breakthroughs in conventional silicon—any could reduce demand for the brute-force approach that today's AI requires. The massive investment in Saline Township bets on a particular trajectory for AI development. If that trajectory shifts, the investment's value shifts with it. The billions of dollars in concrete and copper and steel cannot easily be repurposed.

The community will live with either outcome. Township residents who opposed the project worried about exactly this uncertainty. They questioned whether betting their community's future on a technology whose path is unpredictable was wise. Supporters argued that economic opportunity justifies the risk. Both perspectives remain valid as the facility rises from the farmland.

11.7 ARCHITECTURE FOR AI FACILITIES

The Saline Township facility includes both training and inference capacity, with distinct zones designed for each workload type. This reflects an emerging pattern: facilities optimized for the different demands of each.

Training clusters pack thousands of GPUs into compact spaces where networking constraints can be satisfied. The heat generated per square foot exceeds industrial furnaces. Power delivery for a fifty-megawatt training cluster requires electrical infrastructure at industrial scale. Cooling becomes an architectural driver, not an afterthought.

Inference distributes across many independent servers, enabling different design choices. Power density runs lower; cooling requirements prove less extreme. But scale creates its own challenges. An inference farm serving millions of users might contain tens of thousands of servers with enormous aggregate power consumption.

Training clusters require ultra-low-latency connections between GPUs—microsecond-level responsiveness for gradient exchanges. External networking connects facilities to the broader internet; a large AI facility might require hundreds of gigabits per second of bandwidth. Saline Township benefits from Michigan's position on major fiber routes connecting the East Coast to Chicago.

Commercial data centers promise five nines availability. Achieving this requires systematic redundancy: multiple utility feeds, backup generators, N+1 cooling configurations, component backups throughout. Saline Township incorporates all these elements. The reliability engineering adds cost but protects computing equipment worth hundreds of millions of dollars.

11.8 THE PHYSICAL FOOTPRINT

The Saline Township facility includes three buildings totaling approximately 1.65 million square feet. Each is roughly the size of four football fields.[185] Beyond the data halls lie electrical substations, transformer yards, generator buildings, cooling towers, water treatment facilities, security installations. The cooling towers alone can reject hundreds of megawatts of thermal energy. The complete facility forms a self-contained industrial campus.

The visual and acoustic impacts are substantial. Industrial architecture replaces farmland—a permanent transformation. Cooling systems generate noise, particularly during hot weather. The Saline Township

307

consent agreement includes noise limits; acoustic barriers and directional equipment placement mitigate impacts. Construction traffic during peak buildout creates congestion that rural infrastructure was never designed to handle. Operational traffic runs lower but persists.

11.9 DESIGN PHILOSOPHY DIFFERENCES

Different operators take different approaches reflecting company culture and customer requirements. Amazon Web Services emphasizes operational simplicity and modularity—standardized designs replicated across locations. Microsoft prioritizes sustainability alongside performance. Google focuses on efficiency, achieving industry-leading power usage effectiveness ratios. Meta emphasizes openness through the Open Compute Project. Oracle, the anchor tenant for Saline Township, combines enterprise reliability requirements with cloud efficiency goals.

Operators serving multiple tenants face different priorities. Equinix focuses on interconnection, designing facilities as network meeting points. Digital Realty emphasizes flexibility for diverse tenant requirements. QTS prioritizes speed through modular construction. CoreWeave optimizes specifically for AI density and networking requirements.

11.10 EMERGING TECHNOLOGIES

Several emerging technologies could reshape data center design. Co-packaged optics integrate optical components directly with computing chips, reducing latency, power consumption, and space requirements. High-voltage direct current distribution eliminates conversion steps, achieving efficiency improvements of several percentage points—tens of megawatts of reduced consumption at gigawatt scale. On-chip power conversion, pioneered by Google for its tensor processing units, pushes voltage regulation to the processor itself.

More speculatively: neuromorphic processors designed specifically for neural network operations could dramatically reduce power consumption per unit of useful AI work. Quantum computers, if fault-tolerant systems become practical, could solve problems current AI systems cannot. Neither is ready to replace current infrastructure. Both represent paths to dramatically different requirements—facilities built for GPUs might not serve these alternatives. The transition risk runs in both directions: today's facilities might become obsolete, but facilities designed for speculative future technology might never find appropriate workloads.

11.11 DESIGN FOR RESILIENCE

Climate change is altering the environment in which data centers operate. Heat waves stress cooling systems. The 2021 Texas winter storm revealed vulnerabilities operators had not anticipated—both power disruption and extreme cold.[84] The Midwest faces its own climate risks: severe storms, flooding, heat waves projected to increase in frequency and intensity.

Water availability varies geographically and temporally. The western United States already experiences constraints affecting data center siting. Closed-loop systems like those at Saline Township reduce water vulnerability but do not eliminate it. The power grid faces its own climate vulnerabilities. Backup generators are designed for short outages, not extended failures.

The long planning horizons for data centers—decades of expected operation—must account for conditions that may differ substantially from current patterns. Investment in grid hardening, transmission capacity, and distributed generation can improve reliability. Failure to make such investments leaves vulnerabilities that individual facilities cannot address through their own designs.

11.12 THE ECONOMICS OF INNOVATION

The hyperscalers spend billions annually on research and development—Google's DeepMind, Microsoft Research, Meta's AI Research lab all contribute to data center innovation. Smaller operators depend on innovations that equipment vendors commercialize or that hyperscalers share, creating a technology gap between the largest operators and everyone else.

Even proven innovations face deployment challenges. The Saline Township facility is being built with current technology; upgrading will require future investment. New facilities have more flexibility, creating a vintage effect: newer facilities tend to be more efficient.

Innovation creates stranded asset risk. Technology that improves incrementally maintains facility value; technology that changes fundamentally can strand investments. The billions invested in Saline Township assume the facility will remain valuable for decades. Technology change could reduce that value. Operators manage this risk through design flexibility, financial hedging, and diversification—but no strategy fully eliminates uncertainty.

11.13 THE INNOVATION IMPERATIVE

The rapid evolution of data center technology reflects a deeper truth about AI infrastructure: this is not a settled industry building standardized facilities but a frontier pushing against fundamental constraints.

Power remains the binding constraint. Thermal management determines how much computation can fit in the power available. Every efficiency gain in cooling enables more computation per kilowatt. Every reduction in computing per bit of useful AI output reduces the infrastructure required. These improvements compound. A facility twice as efficient can do the same work with half the power, or twice the work with the same power.

The companies investing in Saline Township and its counterparts across America understand this dynamic. They build not static facilities but platforms for continuous improvement. The initial investment creates capability; ongoing innovation determines how that capability evolves.

This creates a different relationship between technology and community than traditional industrial development. A manufacturing plant builds a product line that may operate for decades with modest changes. A data center builds infrastructure that will transform repeatedly over its lifetime. Communities hosting AI facilities accept no fixed impact. They enter a relationship that will evolve in ways no one can fully predict.

The technologies explored in this chapter—liquid cooling, modular construction, edge computing, experimental frontiers, sustainability innovation—represent the menu of options for that evolution. Some will prove essential; others will fade. The specific mix that Saline Township experiences depends on decisions made by operators, regulators, and communities in the years ahead.

What we can say with confidence: change is coming. The data center being built today is not the data center that will operate in 2035. Communities must ensure that evolution serves their interests as well as the operators'. Operators must innovate rapidly while maintaining the reliability their customers demand. Policymakers must create frameworks that encourage beneficial innovation while protecting communities from its costs.

These challenges define the frontier that data center design is now crossing. The next chapter examines how policy might shape that crossing.

Technology will not solve the fundamental tensions of AI infrastructure development. Cooling innovation does not address community concerns about democratic process. Modular construction does not resolve debates about land use. Nuclear power does not eliminate questions about who benefits and who bears costs. But technology shapes the terms on which these debates occur. Understanding what is possible—and what is not—grounds those debates in reality rather than speculation. The machines in Saline Township will change. The questions the community faces will not.

CHAPTER TWELVE

Future Policy

F IVE YEARS FROM NOW, a township supervisor in rural Ohio will receive a phone call from a company she has never heard of. They want to build a data center on farmland near the Interstate. Two billion dollars in investment. A hundred permanent jobs. Electricity consumption matching a small city. The supervisor will have forty-eight hours to decide whether to meet with them.

She will face the same questions Saline Township faced, with the same limited information. The company will cite the same urgency: competitors in Indiana, other developers circling the same land, decisions that must happen now. And she will wonder whether five years of projects like Saline have taught anyone anything.

The answer depends on choices being made today. Will communities have tools to evaluate proposals? Will utilities have capacity to serve new loads? Will environmental impacts be measured and managed? Will the benefits of AI infrastructure reach beyond the companies that build it? This chapter explores what those choices look like—and what difference they might make.

12.1 THE POLICY ENVIRONMENT

Data center development cuts across multiple policy domains: energy regulation, land use planning, environmental protection, economic development, and increasingly, national security. No single agency controls the entire process. This fragmentation creates opportunities and vulnerabilities in equal measure.

The regulatory framework for data centers evolved when the industry looked different. Before the AI boom, these were modest facilities—tens of megawatts, not gigawatts—that slipped into existing grid infrastructure without major upgrades. Zoning codes filed them under "light industrial" or "technology parks." Environmental reviews focused on backup generator emissions, not water consumption at scale.

The current wave of development exposes gaps in this framework. Gigawatt projects strain transmission systems designed for dispersed loads. Rural zoning codes lack categories for industrial facilities larger than anything the community has ever seen. Environmental reviews that work for individual projects cannot assess cumulative impacts across regions.

These gaps breed uncertainty for everyone. Developers face unpredictable approval timelines and community opposition. Communities confront proposals without tools to evaluate them. Utilities struggle to plan for loads that materialize suddenly and at unprecedented scale. Regulators must decide using frameworks designed for a different era.

––––––

Any policy framework must balance competing interests, all of them legitimate. Developers want fast approvals, predictable regulations, access to power. Communities want jobs, tax revenue, protection from harm. Utilities want cost recovery for infrastructure investments. Environmental advocates want accountability for energy consumption and emissions. Ratepayers want affordable electricity. National security officials want domestic AI capacity.

These interests sometimes align, sometimes collide. A streamlined approval process may deprive communities of information they need. Environmental protections may delay projects that national security requires.

Tax incentives that attract investment may burden ratepayers who receive no direct benefit.

Policy design means choosing among these trade-offs. Different jurisdictions will choose differently, reflecting different values and circumstances. But some choices produce better outcomes than others. The challenge is identifying frameworks that serve broad interests rather than narrow ones.

12.2 GRID MODERNIZATION

The power grid is the physical foundation on which AI infrastructure rests. Every data center proposal depends on available transmission capacity, generation resources, and distribution infrastructure. Grid policy determines what is possible and where.

The most immediate bottleneck is interconnection—the process by which new loads connect to the grid. Utilities and grid operators study proposed connections to ensure they will not destabilize existing service. These studies can take years. Sometimes longer than constructing the data center itself.

The Federal Energy Regulatory Commission has recognized this problem. FERC Order 2023, issued in July 2023, reformed interconnection procedures for generation projects.[97] The order replaced first-come, first-served queues with cluster-based studies that evaluate projects together. It imposed financial commitments to screen out speculative applications and established timelines that, if enforced, would cut study periods significantly.

Similar reforms for large load interconnection are under consideration. Data centers represent a new category of customer, one that existing frameworks never anticipated. A single facility requesting a gigawatt of service creates study complexities that traditional industrial load requests never posed. For context: a gigawatt equals the output of a large power plant.

Some utilities have begun developing specialized tariffs and processes for large technology loads. These "large load riders" establish service terms that account for the unique characteristics of data center demand: high capacity factors, rapid ramp-up, limited flexibility. The approach treats data centers as a distinct customer category requiring distinct treatment.

The risk is that specialized treatment becomes preferential treatment. Data centers that receive faster interconnection than other customers effectively jump the queue. Data centers that receive infrastructure built at ratepayer expense shift costs onto others. Policy must distinguish between appropriate accommodation and inappropriate subsidy.

———

Beyond interconnection, data centers benefit from—and strain—regional transmission infrastructure. A new gigawatt load in rural Michigan draws power from generation sources that may be hundreds of miles away. The wires connecting those sources to the load must have capacity that often does not exist.

FERC Order 1920, issued in 2024, requires regional transmission organizations to plan for long-term needs over twenty-year horizons.[235] Previous planning focused on shorter periods and often underestimated load growth. The new requirements force planners to consider where demand will emerge and build infrastructure to serve it before constraints become acute.

Data center developers have mixed reactions to long-term planning. It creates the transmission capacity they need. But it may also direct investment to regions where planners anticipate growth, not necessarily where developers want to build. Centralized planning substitutes planner judgment for market signals—and developers generally prefer market signals.

The alternative—waiting for congestion to develop and then building infrastructure reactively—produces the delays that frustrate everyone. Long-term planning accepts this trade-off, betting that coordinated investment produces better outcomes than uncoordinated response.

———

Transmission carries power but does not create it. Data centers ultimately depend on generation capacity—power plants that convert fuel or renewable resources into electricity. The data center boom arrives as the generation fleet is transforming: coal plants retiring, renewable capacity expanding.

This transformation raises adequacy concerns. Renewable generation—primarily solar and wind—produces power intermittently. Solar panels generate nothing at night. Wind turbines sit idle on calm days. Meeting data center demand, which runs twenty-four hours a day, requires either dispatchable generation that fires on demand or storage that banks intermittent production.

Some operators address adequacy through direct arrangements with generators. Microsoft's deal with Constellation Energy to restart the Three Mile Island nuclear plant is the most prominent example.[229] The arrangement secures dedicated generation for Microsoft's data centers, insulating the company from grid constraints.

Not all operators can make such arrangements. Smaller players lack the scale to justify dedicated plants, and even large operators face a stubborn reality: available generation cannot meet all demand simultaneously. The generation fleet must expand to serve data centers—renewable or otherwise.

Policy can accelerate this expansion or impede it. Permitting reform for generation projects—particularly transmission interconnection for renewable facilities—would increase supply. Carbon pricing would internalize environmental costs, guiding investment toward lower-emission sources. Clean energy mandates can accelerate the transition, though they may also constrain supply during the transition period.

12.3 ENVIRONMENTAL REGULATION

Data centers consume electricity, water, and land. They emit carbon through electricity consumption, generate noise from cooling equipment, and produce electronic waste as hardware cycles through obsolescence. Environmental regulation addresses these impacts, though current frameworks leave significant gaps.

Every kilowatt-hour a data center consumes comes from somewhere. The carbon intensity of that electricity depends on the generation mix serving the facility: coal plants emit more than natural gas, which emits more than solar or nuclear.

Corporate sustainability commitments typically claim credit for renewable energy through power purchase agreements and renewable energy certificates. A data center in Michigan might purchase credits from a wind farm in Kansas. On paper, the company achieves carbon neutrality. In practice, the Michigan facility draws electrons from whatever generation happens to be running.

This accounting convention has drawn increasing criticism. Environmental advocates argue that true decarbonization requires "additionality"—new renewable generation that would not exist without the data center purchase—and "hourly matching"—renewable production that actually occurs when the data center consumes power. Annual matching with credits from distant sources does not ensure that grid operations actually change.

Some companies are moving toward more rigorous accounting. Google has committed to operating on carbon-free energy twenty-four hours a day, seven days a week, at all its data centers by 2030.[22] Achieving this requires not just purchasing credits but actually matching consumption to clean generation in real time, across every location.

Policy could mandate similar rigor. Requirements that large industrial loads demonstrate hourly matching with local clean generation would transform how data centers source power. Such requirements would also increase costs and complexity, potentially slowing deployment.

———

Water use receives less attention than carbon but increasingly shapes where data centers can go. Traditional evaporative cooling systems consume millions of gallons annually. In water-stressed regions, this consumption competes with agricultural, municipal, and environmental needs.

Some jurisdictions are imposing restrictions. Arizona now requires new developments to demonstrate sustainable water supplies, affecting data center proposals in the Phoenix metropolitan area. Utah has considered limits on water use per megawatt of capacity. These local rules reflect regional water scarcity that varies dramatically across the country.

Federal water policy has not directly addressed data centers, but general environmental requirements affect water-intensive projects. The National Environmental Policy Act requires environmental assessment of federal actions, which can include permits for projects affecting water resources. State laws impose similar requirements.

The most effective water policy may be technological requirements rather than consumption limits. Mandating closed-loop cooling systems, requiring reclaimed water, or incentivizing waterless technologies would reduce consumption while allowing development to proceed. These approaches address the impact directly rather than through permit-by-permit negotiations.

———

Data centers transform land. Agricultural fields become industrial facilities. Rural communities become technology hubs. Open horizons acquire industrial character. Traditional environmental review evaluates projects individually, but the cumulative impact of multiple projects in a region can exceed the sum of individual effects.

Northern Virginia illustrates what cumulative impact looks like. Loudoun County hosts the densest concentration of data centers in the world, with facilities covering thousands of acres.[182] No single project transformed the county, but the accumulation over two decades changed its character fundamentally. Traffic, noise, visual impact, and infrastructure demands compound across projects in ways that individual reviews never anticipated.

Cumulative impact assessment requires regional planning that most American jurisdictions do not practice. Local zoning operates parcel by parcel. Environmental review evaluates projects against existing conditions, not projected build-out. The result: development that proceeds incrementally toward outcomes no one explicitly chose.

Some regions are attempting regional approaches. Northern Virginia jurisdictions have discussed coordinated data center policy, though agreement remains elusive. Pennsylvania has tried to channel growth toward areas with adequate infrastructure. These efforts face a fundamental challenge: regional coordination requires giving up local autonomy.

12.4 INCENTIVES

Practically every state in America offers incentives to attract data center investment. Tax abatements, exemptions, credits, grants—all transfer value from existing residents and businesses to incoming data centers. The transfers are justified as investments that produce returns through jobs, tax revenue, and economic activity. Whether the returns justify the transfers is increasingly debated.

State competition for data centers has intensified to the point of absurdity. Michigan's SB 237 exempts qualifying data centers from sales and use tax through 2050. Potentially twenty-five years of forgone revenue.[80] Georgia's incentive package can cut property tax burdens by ninety percent. Mississippi offers one hundred percent property tax abatements with no clawback provisions—no contractual terms requiring companies to return incentives if they miss job targets.

This competition transfers bargaining power from states to developers. A company shopping for sites can play jurisdictions against each other, extracting larger concessions than any single jurisdiction would offer alone. The result approaches a prisoner's dilemma: each state offers more to avoid losing investment, and collectively, states give away more than necessary.

Economic research on incentive effectiveness is mixed. Some studies find that incentives influence location decisions; others find that companies would have invested anyway, and incentives merely transferred value. The mixed results suggest that incentives matter sometimes, for some projects, but cannot be presumed effective in general.

What is clear is that the current arms race exceeds any plausible efficiency justification. The Saline Township subsidy—two hundred thousand dollars per job—illustrates the scale of current competition. Few other industries receive subsidies at this level. The data center industry's ability to extract them reflects bargaining power, not economic logic.

One response to the incentive problem is community benefit agreements—negotiated packages that specify what communities receive in exchange for hosting data centers. Rather than general tax incentives, CBAs target specific community needs: infrastructure improvements, job training programs, environmental protections, direct payments.

Several states have begun requiring or encouraging community benefit agreements for large data centers. Under such frameworks, projects exceeding certain thresholds negotiate agreements with local governments: minimum wages for construction workers, local hiring commitments, environmental performance standards, community investment funds. Oregon, Washington, and Illinois have each considered or implemented variations.

The Saline Township consent agreement resembles a CBA, though it emerged from litigation rather than policy.[56] The agreement includes provisions for fire department funding, well monitoring, noise limits, and farmland preservation. These specific commitments address community concerns more directly than general tax incentives would.

CBAs shift the negotiating dynamic. Instead of communities competing to offer incentives, developers must compete to offer benefits. The shift does not eliminate bargaining—developers still hold the threat of locating elsewhere—but it changes what gets bargained over. Communities get specific commitments instead of hoping general economic activity produces benefits.

———

Data centers benefit from grid infrastructure that existing ratepayers funded. Substations built to serve residential neighborhoods. Transmission lines constructed decades ago. Generating plants financed through utility rates. All become resources that data centers draw upon. The question is whether data centers should pay their share.

Texas confronted this question directly. Senate Bill 6, passed in 2025, requires large industrial customers—including data centers—to fund grid infrastructure they benefit from.[91] Developers cannot connect to the grid without paying for substation upgrades, transmission reinforcement, and other improvements their load requires.

The policy represents a shift from historical practice, where infrastructure costs spread across all ratepayers. Under the old model, residential customers effectively subsidized industrial development by sharing costs that benefited industry disproportionately. The new model allocates costs to beneficiaries.

Pennsylvania is considering similar legislation. The state's data center boom has strained grid infrastructure, and ratepayer advocates have raised concerns about cost shifting. Proposed legislation would require cost-responsibility assessments for large new loads, ensuring that developers fund necessary upgrades rather than ratepayers.

Ratepayer protection creates tension with incentive competition. A state that requires developers to pay infrastructure costs becomes less attractive than states that spread costs across ratepayers. Protection benefits existing residents but may discourage investment that would create new jobs. Policymakers must weigh the trade-off.

12.5 THREE REGULATORY SCENARIOS

The policy choices described above can combine in different ways, producing different regulatory environments. Consider three scenarios representing different approaches to governance.

In the first scenario, concern about environmental and community impacts drives stricter regulation. States impose water consumption limits, hourly carbon matching requirements, cumulative impact assessments. Community benefit agreements become mandatory. Ratepayer protection shifts infrastructure costs to developers.

Under tightening regulation, data center development slows. Costs rise as developers pay for infrastructure and environmental mitigation. Some projects become uneconomical; developers pursue facilities in countries with lighter regulation.

The benefits accrue to communities and the environment. Projects that proceed deliver genuine local benefits, not promises. Carbon accounting reflects actual emissions, not paper credits. Ratepayers avoid subsidizing corporate infrastructure.

The costs appear in reduced investment and slower AI development. Companies that cannot build in America build elsewhere—potentially in countries with weaker environmental standards. Domestic AI capabilities may fall behind international competitors.

In the second scenario, current trends continue without major policy changes. States compete with incentives. Environmental review proceeds project by project. Grid planning remains reactive. Communities negotiate individually with developers who hold information and bargaining advantages.

Development continues at current pace with uneven outcomes. Some communities benefit substantially; others bear costs without commensurate benefits. Regional variation reflects local politics: data-center-friendly jurisdictions attract investment, skeptical ones deter it. Cumulative impacts emerge over time but get addressed only when they become acute.

The advantage is avoiding disruptive change. The disadvantage is that emerging problems do not get addressed until they become crises.

In the third scenario, federal policy establishes a national framework for AI infrastructure while preserving state flexibility. Grid modernization accelerates through coordinated planning. Environmental standards set minimums that states can exceed but not undercut. Community benefit requirements ensure broad distribution of gains.

Strategic coordination treats AI infrastructure as a national priority—analogous to how earlier eras treated railroads, highways, or telecommunications. Federal investment supports grid expansion in strategic locations. National standards prevent a race to the bottom on environmental and labor protections.

This scenario requires political consensus that does not currently exist. States that prefer autonomy would resist. Developers that prefer negotiating with individual jurisdictions would resist. But it offers the best prospect for outcomes that serve broad interests: accelerated deployment, environmental accountability, community protection, and American competitiveness in AI.

12.6 SALINE REVISITED

Imagine Saline Township in January 2031. Five years have passed since Governor Whitmer announced the project. The facility is operational. What has the community experienced?

The developer promised two thousand five hundred construction jobs. Maybe these materialized, though not all went to local workers. Maybe union rules required project labor agreements, and qualified workers came from across the region. Maybe local hotels and restaurants thrived during construction; some expanded to meet demand.

The developer promised four hundred fifty permanent jobs. Maybe this proved roughly accurate, though job quality varied. Security guards and maintenance workers earn modest wages; engineers and technicians earn considerably more. Maybe the median salary falls below initial projections—lower-wage positions make up a larger share than anticipated.

The developer promised tax revenue that would quadruple township receipts. Maybe the actual increase fell short—partly because state exemptions reduced taxable value, partly because assessment disputes dragged on. Maybe township revenue roughly doubled. Meaningful, but less than promised.

———

Opponents worried about groundwater. The consent agreement required monitoring and mitigation. Maybe early monitoring showed no significant impact on neighboring wells, though the monitoring period is too short for confident conclusions. Maybe the closed-loop cooling system performed as designed.

Opponents worried about electricity rates. Maybe DTE Energy's rate cases—regulatory proceedings where utilities request permission to change prices—included infrastructure costs partially attributable to data center loads. Maybe residential rates rose, though isolating the data center contribution from other factors proves difficult.

Opponents worried about community character. Maybe the facility is visible from Michigan Avenue, though landscaping and setbacks soften the visual impact. Maybe traffic spiked during construction and remains elevated for operations, though not to levels that overwhelm local roads. The rural character that long-time residents valued would have changed— whether for better or worse would depend on who you ask.

———

Some outcomes neither side anticipated. Maybe Oracle's presence attracted auxiliary businesses: equipment suppliers, specialty contractors, consulting firms. Maybe a cluster of data-center-adjacent companies established offices in the region, creating employment beyond the facility itself.

Maybe the University of Michigan expanded computer science programs, partly in response to industry demand. Maybe students who might have left Michigan for jobs in California or Virginia now consider local opportunities.

Maybe property values in the immediate vicinity of the facility declined, while values elsewhere in the county rose. Maybe homeowners adjacent to the data center lost; homeowners farther away gained from regional economic growth.

––––––

Five years after approval, maybe Saline Township's experience defies simple summary. Maybe the project delivered some promised benefits but not all. Maybe it produced some feared impacts but not the worst. The community would be different—wealthier in aggregate, more industrial in character, more connected to global technology markets.

Whether this was "worth it" would depend on values that reasonable people hold differently. Residents who wanted economic development would point to jobs and tax revenue. Residents who wanted to preserve rural character would point to what was lost. Neither side would be entirely wrong about what happened. They valued different things.

The experience offers lessons for other communities facing similar decisions. Specific commitments in consent agreements matter more than general promises. Monitoring provisions can verify claims, but require enforcement. Regional economic effects extend beyond the facility itself. And five years is not long enough to evaluate impacts that may take decades to fully emerge.

12.7 POLICY RECOMMENDATIONS

Drawing on the analysis throughout this chapter, we can identify policy approaches that would improve outcomes for communities, operators, and the nation.

Federal policy can establish frameworks that states alone cannot.

The transmission system is the binding constraint on AI infrastructure. Grid modernization could accelerate if the federal government shared costs for interstate transmission projects that no single state will fund alone. Interconnection studies could be capped at twenty-four months for technically ready projects, ending the multi-year queues that delay development. Dedicated funding could support high-voltage corridors linking renewable-rich regions to data center clusters. Pre-approved corridors along Interstate highway rights-of-way would eliminate years of route disputes. These mechanisms exist in fragmentary form today; consolidating them into coherent policy would accelerate what the current patchwork impedes.

Federal environmental standards would prevent a race to the bottom among competing states. Water efficiency requirements could limit consumption to 1.0 gallons per kilowatt-hour in water-stressed regions, forcing closed-loop cooling rather than evaporative systems that drain aquifers. Elsewhere, a 1.5-gallon limit would allow some evaporative cooling while still constraining consumption. Mandatory hourly carbon accounting by 2030 would replace the current practice of annual matching through distant credits. Operators would need to demonstrate that clean generation actually runs when their facilities consume power.

Cumulative impact assessments, triggered when regional data center capacity exceeds five gigawatts, would force planners to consider what happens when multiple facilities arrive in the same region. Public reporting of power usage effectiveness, water usage effectiveness, and carbon intensity would let communities and regulators compare operator claims against actual performance.

AI infrastructure is a strategic asset, but treating every project as a national security priority distorts local decision-making. Clear guidelines about which projects warrant expedited treatment and which should proceed through normal processes would improve both security and democracy.

State policy shapes the immediate environment for data center development.

States collectively lose billions in foregone revenue through data center incentives. Reforming incentive competition would require independent analysts to verify that projected benefits exceed costs before any state offers more than ten million dollars. Clawback provisions would recover half of incentive value if employment targets miss by more than twenty percent—real consequences for broken promises, not symbolic penalties. Sunset provisions could terminate exemptions after fifteen years, forcing explicit renewal rather than open-ended commitments that no current official will have to defend. An interstate compact could cap total incentives at fifty thousand dollars per job created, preventing the escalating bidding wars that transfer value from states to developers. Such a cap would still allow meaningful incentives while limiting the most egregious giveaways.

Required community benefit agreements would shift negotiating dynamics toward communities. Any project exceeding one hundred megawatts would trigger mandatory negotiation. Minimum hiring requirements—a quarter of construction workers and half of permanent employees from within fifty miles—would ensure that local residents benefit from employment rather than watching jobs flow to imported labor. A community investment fund set at half a percent of project cost would translate billion-dollar investments into meaningful local resources: a seven-billion-dollar project would contribute thirty-five million dollars to community priorities. Annual third-party audits with public reporting would verify compliance, giving communities recourse when operators fail to deliver.

Ratepayer protection requires that infrastructure costs be borne by beneficiaries, not spread across customers who receive no benefit. Cost-responsibility assessments for loads exceeding fifty megawatts would ensure that data centers pay for substation upgrades, transmission reinforcement, and other improvements their demand requires. Rate class separation would prevent utilities from recovering data center infrastructure costs from residential customers—the quiet subsidy that currently shifts hundreds of millions of dollars from homeowners to corporations. Interconnection costs could be capped at five years' projected revenue, preventing unlimited obligations while ensuring developers fund the infrastructure they need. The difference could be hundreds of dollars annually on residential bills in regions with heavy data center concentration.

Current policy treats all data centers identically regardless of their relationship to existing infrastructure. A smarter approach would incentivize urban integration, rewarding facilities that solve urban problems rather than create rural ones. Tax incentives could favor facilities that connect to district heating networks, repurpose aging infrastructure, or locate on genuinely underutilized urban land. The brownfield premium Michigan attempted in 2025 pointed in the right direction but applied to gigawatt-scale facilities that cannot practically locate in urban areas. Incentives scaled to facility size—stronger benefits for smaller urban-integrated projects—would encourage the Deep Green model rather than the Saline model.

––––––

Local decisions ultimately determine whether projects proceed.

Individual townships and counties lack negotiating power against sophisticated developers. Regional coordination strengthens local positions through joint planning, shared analysis, and coordinated negotiation. Northern Virginia's data center alley jurisdictions demonstrate both the difficulty and necessity of coordination.

Communities cannot evaluate proposals they do not understand. Disclosure requirements—power consumption, water use, employment projections, environmental impacts—should precede any approval. Claims should be verifiable and verified.

Urgency benefits developers, not communities. Projects that cannot wait for adequate review may not deserve approval. The costs of approving bad projects exceed the costs of delaying good ones.

12.8 INTERNATIONAL DIMENSIONS

AI infrastructure policy is not purely domestic. International competition shapes federal decisions, while coordination with allies influences how standards develop.

Strategic competition with China runs through American AI policy, though the policy instruments shift with administrations. The October 2022 export controls restricted advanced chips; the January 2025 AI Diffusion Rule attempted global performance thresholds; the May 2025 rescission reversed course; the December 2025 H200 approval created a revenue-sharing model that earlier policymakers might have considered unthinkable.[47]

The rapid shifts create infrastructure planning challenges. Companies building facilities in 2025 and 2026 cannot know what export regime will govern chip sales in 2028 or 2030. If restrictions tighten, Chinese demand for American AI services might grow as domestic capacity proves inadequate. If restrictions loosen further, Chinese data centers might compete directly for the same workloads. The strategic calculus depends on political variables that no financial model can capture.

The infrastructure implications cut both ways. Tight restrictions buy the United States a window of advantage but accelerate Chinese self-sufficiency efforts. Loose restrictions preserve American market share but may narrow the capability gap. The December 2025 policy suggests the current administration prioritizes revenue capture over maximum separation. Future administrations may choose differently.

Export controls also affect domestic infrastructure decisions, though the mechanism has evolved. Under maximum restriction, the urgency to build reflected competitive pressure—the sense that America must race to deploy before China catches up. Under revenue-sharing, the calculus shifts: American infrastructure competes with Chinese alternatives on commercial terms, not just strategic ones. The Trump administration's December 2025 policy effectively concedes that some Chinese AI development will proceed, seeking instead to profit from it.

Some observers question whether the competitive framing remains accurate. China has substantial AI capabilities despite years of export controls. DeepSeek demonstrated frontier-competitive performance with dramatically less compute. Alternative chip architectures continue emerging. The race metaphor may overstate urgency while understating the domestic costs of hasty infrastructure deployment. Others counter that the revenue-sharing model acknowledges competitive reality without abandoning strategic position—that monetizing what cannot be prevented beats prohibiting what cannot be enforced.

The United States is not alone in building AI infrastructure. European countries, Japan, South Korea, and other allies are also investing in domestic capacity. Coordination among allies could reduce duplication and ensure that democratic nations collectively maintain AI advantages.

Current coordination is limited. Each country pursues independent infrastructure strategies. European data centers face different regulatory environments than American ones. Japanese facilities emphasize different design priorities. This variation creates inefficiency—but also experimentation that may produce useful innovations.

Common standards could help. If democratic nations agreed on sustainability requirements, data protection standards, and security protocols, the resulting infrastructure would be more interoperable. Companies operating across borders would face consistent expectations. But achieving such agreement requires sustained diplomatic effort that has not yet materialized, to put it lightly.

The Saline Township project illustrates how local infrastructure decisions remain. Oracle's choice to build in Michigan reflected Michigan-specific factors: power availability, tax incentives, community receptivity. International coordination would not change these local dynamics, though it might shape the broader context within which local decisions occur.

Nations increasingly view AI infrastructure as a sovereignty issue. Depending on foreign cloud providers for AI capabilities creates vulnerability—if geopolitical tensions escalate, access to essential computing resources could be disrupted.

This concern drives domestic investment even when economic logic might favor importing services. France, Germany, and other European nations have launched sovereign cloud initiatives. Middle Eastern countries are building domestic AI capacity despite limited technical workforces. India has articulated ambitions for AI self-sufficiency.

The sovereignty trend reinforces global buildout. Rather than concentrating capacity in a few efficient locations, infrastructure disperses across national markets. Each nation wants domestic capability, even when it costs more than imports would.

For American policy, sovereignty concerns cut both ways. American infrastructure serves American sovereignty but may also serve foreign customers whose home countries lack capacity. If those customers later develop domestic alternatives, American facilities face reduced demand. The stranded asset risk includes not just technology change but geopolitical change.

12.9 WORKFORCE AND LABOR

AI infrastructure creates jobs, but the nature of those jobs shapes who benefits from the buildout.

Construction employment is substantial but temporary. These jobs pay well—union rates for skilled trades—and draw workers from across the region. When the facility is complete, workers move to other projects. The jobs boost local economies during construction but do not last. Communities hoping for permanent job creation must look beyond the construction phase.

Union requirements in the consent agreement ensured that construction jobs met prevailing wage standards. This protects workers from exploitation but also increases project costs. The trade-off reflects values: prioritizing worker welfare over pure cost minimization.

———

Permanent data center employment is modest relative to investment. A billion-dollar facility might employ fewer than one hundred people in ongoing operations. Even Saline Township's full buildout, at the high end of the range, employs far fewer than the construction workforce.

Operations jobs span a range of skills. Security personnel earn modest wages. Technicians maintaining equipment earn more. Engineers managing complex systems earn substantially more. The median salary depends on the job mix—which developers may not disclose during project negotiations.

Automation is advancing. Tasks that once required human attention increasingly happen automatically. AI systems monitor other AI systems. Remote operations centers manage multiple facilities. The trajectory points toward fewer workers per megawatt over time.

———

Communities hoping to capture data center employment must develop appropriate skills in their workforce. Electricians, HVAC technicians, and network engineers are in demand. Training programs that produce qualified workers can help residents access these jobs.

Some operators partner with local community colleges and technical schools. These partnerships create training pathways for residents who might not otherwise reach technology sector employment. The programs benefit operators, who need workers, and communities, who want good jobs for residents.

The Saline Township consent agreement includes workforce development provisions, though specifics remain limited. More comprehensive requirements might ensure that local residents benefit from employment opportunities rather than seeing jobs flow to workers from elsewhere.

12.10 DEMOCRATIC PROCESS

Beyond substantive policy choices, process matters. How communities participate in data center decisions shapes whether outcomes reflect community values.

Developers possess information that communities do not: project economics, customer commitments, technical specifications, timeline pressures. This asymmetry advantages developers in negotiations.

The Saline Township experience illustrated the problem. Board members asked basic questions that developers could not or would not answer. Who is the tenant? The developer would not say. What are the water consumption estimates? Vague assurances only. Communities were asked to approve projects whose details remained hidden.

Better policy would mandate disclosure. Before public hearings, developers would provide comprehensive information about power consumption, water use, employment projections, and environmental impacts. Claims would be verifiable. Communities would evaluate proposals on facts rather than promises.

335

Developers create urgency. Competitors are circling. Other sites are available. Decisions must happen now. This pressure limits community deliberation and advantages developers who benefit from hasty approvals.

The timeline pressure is often genuine. Data center demand is real, and operators who cannot secure sites lose customers to competitors. But urgency is also tactical. Developers know that extended review creates opportunities for opposition to organize. Fast approval forecloses scrutiny.

Policy could counter this. Mandatory review periods would establish that data center decisions take time regardless of developer preferences. Communities would have opportunity for analysis and deliberation. Developers who cannot wait find other sites; those who want specific locations accommodate community timelines.

When Saline Township rejected the initial proposal, Related Digital filed suit within days. The lawsuit created financial pressure the township could not sustain—defending would cost more than the township could afford. Settlement became inevitable.

This dynamic applies across jurisdictions. Developers with deep pockets outlast communities with limited resources. Legal threats become negotiating tools. The threat of litigation shapes outcomes even when cases never reach trial.

Equalizing legal resources is difficult. Some states provide legal assistance to localities facing developer lawsuits. Public interest law organizations sometimes represent communities. But the asymmetry is fundamental. Policy cannot fully address the imbalance between billion-dollar corporations and million-dollar townships.

Local governments make local decisions. This fragmentation empowers developers who play jurisdictions against each other. If one township says no, an adjacent one might say yes. The threat of losing investment to neighbors constrains what any single jurisdiction can demand.

Regional coordination could address this. If jurisdictions agreed on common standards, developers could not exploit differences. Minimum requirements applied across regions would establish floors no jurisdiction could undercut.

But coordination is difficult. Local governments guard their autonomy jealously. Agreements require giving up control over local decisions. The same fragmentation that empowers developers prevents the coordination that might constrain them.

12.11 THE STAKES

Policy decisions made in the next few years will shape AI infrastructure development for decades. The choices are consequential.

Weak policy allows development to proceed on terms that benefit developers without adequate protection for communities or the environment. Strong policy ensures that the AI buildout delivers broad benefits while managing its costs. Strategic policy positions America for long-term competitiveness while preserving the values that define American governance. The choice among these approaches will shape outcomes for decades.

The trillion-dollar investment in AI infrastructure is coming regardless of policy choices. The question is whether that investment strengthens or strains American communities, whether its benefits are broadly shared or narrowly captured, whether the political system can manage transformation at this scale.

Saline Township offers a preview. The community negotiated under pressure, with imperfect information, facing an opponent with vastly greater resources. The outcome was neither victory nor defeat—a compromise that left neither side fully satisfied. Most communities facing similar proposals will experience similar dynamics.

Policy cannot eliminate these tensions. What it can do is shift the terms on which they are resolved: better information, clearer standards, fairer distribution of costs and benefits, accountability for promises made. Modest improvements compound across hundreds of projects to produce meaningfully different national outcomes.

The Ohio township supervisor who receives that phone call in five years will face the same decision Saline Township faced. Whether she has better tools to make it depends on choices made today.

12.12 INTERNATIONAL LESSONS

The United States is not the only country building AI infrastructure. Other nations have developed approaches that offer lessons—both positive and negative—for American policy.

The European Union takes a more regulatory approach to data center development. Environmental impact assessments run more comprehensive, energy efficiency requirements more stringent, community consultation more extensive.

Nordic countries have developed district heating integration that captures waste heat from data centers. These systems require coordination between operators, utilities, and municipal governments that American frameworks rarely achieve. The efficiency gains are substantial—but they require policy infrastructure that must be built before individual projects can benefit.

Ireland has become a cautionary tale. Favorable tax policies attracted massive investment, but data centers now consume roughly twenty-two percent of Irish electricity.[236] The government has imposed restrictions on new development. Northern Virginia already shows signs of following this pattern: grid capacity limits constraining development despite strong demand.

338

Singapore treats data center capacity as national infrastructure, allocating power through a managed process that keeps development pace with demand without overwhelming the grid. Japan emphasizes disaster resilience, with design standards exceeding American requirements for seismic resistance.

China has directed development toward regions with abundant renewable energy, demonstrating that policy can shape data center geography—though the mechanisms available differ from those in market democracies.

American policy can learn from international experience without copying it directly. Nordic waste heat integration shows what coordination can achieve. Ireland's concentration problem warns against letting development outpace infrastructure. Singapore demonstrates alternatives to pure market-driven siting.

The challenge is adapting these lessons to American institutions. Federal structure limits national coordination. Property rights constrain land use regulation. Policy ideas that work elsewhere must be translated into forms compatible with American governance.

12.13   Emerging Policy Tools

Beyond the established approaches discussed earlier, several emerging tools could reshape how data center development is governed.

If the United States implements carbon pricing, carbon border adjustments would prevent carbon-intensive production from simply moving abroad. Whether through a carbon tax or cap-and-trade system, the principle is the same. Applied to data centers, such adjustments would impose costs on computing services imported from regions with higher carbon intensity.

The European Union has implemented a Carbon Border Adjustment Mechanism that applies to certain industrial products. Extending similar mechanisms to digital services would be technically challenging but not impossible. Computation powering AI services could be traced to data center locations, and associated carbon intensities assessed.

Such mechanisms would create incentives for clean energy at data centers regardless of location. American facilities powered by renewables would gain advantage over foreign facilities powered by coal. Climate and economic objectives would align.

———

Direct federal investment in grid infrastructure could accelerate transmission expansion beyond what current policy achieves. The Bipartisan Infrastructure Law and Inflation Reduction Act included significant funding for grid modernization.[134] Additional investment could push the transition faster still.

Infrastructure investment raises questions about who benefits and who pays. If federal funds enable data centers that serve private companies, should those companies share the costs? The answer depends on whether data center infrastructure is treated as general-purpose investment benefiting the economy broadly, or as targeted subsidy benefiting specific firms.

———

Federal or state preemption of local zoning could accelerate data center development by preventing local opposition from blocking projects. Such preemption already exists for other infrastructure types, including telecommunications facilities under certain conditions.

The trade-off is stark. Preemption enables development that local opposition would prevent, but it also overrides local democratic processes that give communities voice in what happens within their boundaries. The same preemption that enables beneficial projects enables harmful ones.

Targeted preemption might split the difference. Projects meeting certain criteria—environmental standards, community benefit requirements, infrastructure investment thresholds—could qualify for expedited approval. Projects failing to meet criteria remain subject to local review. The criteria become the battleground for policy debate.

Some have proposed treating data centers as public utilities subject to rate regulation and service obligations. The argument: AI computing has become essential infrastructure, comparable to electricity or telecommunications. Private companies should not control access to capability that society increasingly depends upon.

Utility regulation would mark a dramatic departure from current practice. Data centers are private facilities serving private customers. Imposing public utility obligations would require new legal frameworks and regulatory capacity. The transition costs would be substantial.

But the comparison to electricity and telecommunications is instructive. Both were once private services that became public utilities as their essential nature became clear. AI computing may follow a similar path if it becomes sufficiently central to economic and social life.

12.14 THE LONG VIEW

Policy evolves over time. The frameworks governing data centers today will change as experience accumulates and circumstances shift.

The projects approved today will produce outcomes that inform future policy. If Saline Township flourishes, similar projects face easier approval. If problems emerge, restrictions tighten. The lived experience of communities shapes what future communities are willing to accept.

This learning process is valuable but slow. Years pass between project approval and measurable outcomes. The lag between action and feedback limits how quickly policy can adapt.

Accelerating learning requires better monitoring. If communities tracked outcomes systematically—employment, tax revenue, environmental impacts—lessons would emerge faster. The industry's rapid growth makes timely learning especially valuable.

The institutions governing data center development are adapting. Utility commissions are developing expertise in large load management. Local governments are learning from peer experiences. Industry associations are codifying practices.

But adaptation lags need. The pace of investment outstrips institutional development. Utilities approve projects before fully understanding grid implications. Local governments negotiate agreements before knowing what to demand.

Closing this gap requires investment in institutional capacity: training for local officials, technical assistance programs, information sharing among jurisdictions. The costs are modest relative to the stakes, yet receive limited attention.

Policy responds to political cycles. Administrations change. Majorities shift. The priorities that shape data center policy today may not be tomorrow's priorities.

The current policy environment favors development. Federal policy emphasizes AI competitiveness; state policy emphasizes economic development. These emphases could reverse if political winds shift: environ-mental concerns, ratepayer backlash, or community opposition could re-shape political calculations.

Operators and investors face this political uncertainty alongside tech-nology and market uncertainty. Projects with decade-long timelines must account for political environments that may change multiple times dur-ing development and operation.

The next chapter examines where these policy choices lead. We con-struct three scenarios for AI infrastructure through 2030—exploring how different assumptions about technology, policy, and economics produce different futures for communities like Saline Township.

CHAPTER THIRTEEN

Our Future

I MAGINE IT IS JANUARY 2030. Four years have passed since the first construction equipment arrived at the Saline Township site. The facility that seemed like a distant vision is now part of daily life. What does that life look like?

The honest answer: we do not know. AI infrastructure depends on variables no one can predict with confidence. Will capabilities keep improving at current rates, or will progress slow? Will electricity demand match projections, exceed them, or fall short? Will policy tighten or loosen? Will some breakthrough reshape what computing requires?

This chapter explores three scenarios for how AI infrastructure might evolve through 2030. Each represents a plausible future under different assumptions about technology, economics, and policy. None is a prediction. All are attempts to think through what different futures would mean for communities like Saline Township and for the nation they belong to.

Scenario planning prepares us for possibility, not certainty. The three scenarios span continued exponential growth, stagnation and stranded assets, and transformational technological change. Reality will likely combine elements of each.

13.1 THE FRAMEWORK

Before examining specific scenarios, we need a framework for thinking about what drives them. Three forces shape AI infrastructure: technology, economics, and policy.

Sarah thinks about scenarios every day. In the PJM control room, grid operators cannot afford to plan for a single future. "We run contingencies constantly," she says during a break between shift handoffs. "What if this generator trips? What if demand spikes? What if a transmission line goes down? The grid only works because we've already thought through the alternatives."

The discipline extends to longer-term planning. *Sarah* sits on a regional planning committee that develops fifteen-year forecasts. "When I started, the forecasts were mostly about retirements—coal plants closing, demand flattening. Now the forecasts are about load growth we've never seen. Data centers are the single biggest variable. If they build what they've announced, we need to double transmission capacity. If they don't, we've overbuilt."

Her uncertainty mirrors the uncertainty facing every actor in this space. The framework that follows reflects the same disciplined thinking about alternatives.

AI capabilities have improved rapidly, but the rate and duration of improvement remain uncertain. *Technology* shapes what is possible.

Model efficiency shapes what is possible. How much computation achieves a given level of AI performance? If efficiency improves faster than demand grows, infrastructure needs plateau or decline. If efficiency gains lag capability growth, infrastructure needs explode. Current trends show both capability and demand growing faster than efficiency.[17]

Each generation of AI accelerators delivers more computation per watt. NVIDIA's roadmap promises substantial gains—Blackwell delivers roughly five times higher throughput per megawatt than Hopper.[24] If this continues, the same AI tasks require less infrastructure over time. If chip improvements slow, whether from physical limits or manufacturing challenges, infrastructure requirements grow.

Alternative architectures could change everything. Today's AI infrastructure runs on graphics processing units optimized for matrix multiplication. Alternative approaches—quantum computing, neuromorphic processors, photonic computing—could change hardware requirements fundamentally. None is commercially ready. But breakthroughs could arrive faster than expected.

Technology determines what is possible; economics determines what is profitable.

AI demand determines whether the infrastructure gets used. How much will customers pay for AI services, and how does that demand grow over time? Current enthusiasm has driven enormous investment, but enthusiasm can wane. Enterprise adoption, consumer applications, and government use all contribute. If any category disappoints, investment may slow.

Capital availability matters as much as technology. AI infrastructure requires massive capital: billions per project, trillions across the industry. Interest rates, risk appetite, and competition for investment all affect how much flows to data centers. The low-interest environment that fueled the initial boom may not persist.

Electricity constitutes a major operating expense. If energy costs rise—from fuel prices, carbon charges, or infrastructure constraints—project economics shift. Locations with cheap power become more attractive; expensive locations become untenable.

Policy shapes the environment within which technology and economics operate.

Environmental regulation affects project economics. Stricter requirements for carbon emissions, water consumption, or land use increase costs and slow deployment. Looser requirements accelerate it. Current trends suggest incremental tightening but no dramatic shifts.

Grid policy determines how quickly transmission and generation infrastructure can expand. Permitting rules, cost allocation, and planning horizons determine how much load the grid can serve and where.

International competition shapes federal priorities. If other countries build AI infrastructure more aggressively, the United States faces pressure to accelerate domestic deployment regardless of local costs. Strategic competition with China shapes federal policy above all.

13.2 EXPONENTIAL GROWTH

In this scenario, AI capabilities continue improving rapidly, demand exceeds projections, and infrastructure deployment accelerates.

The scenario assumes AI model capabilities keep improving at rates comparable to 2020–2025, with each generation delivering meaningfully greater performance. Enterprise customers double their AI budgets. Consumers treat AI assistants like smartphones—tools they cannot imagine abandoning. Capital remains abundant, with institutional investors competing to fund data center projects. Policy accommodates growth, with environmental constraints modest enough to avoid slowing deployment. Grid infrastructure expands through expedited permitting, with utilities prioritizing data center interconnection.

This is the future data center developers are betting on. Current trends continuing without major disruption.

———

Under continued exponential growth, the United States in 2030 hosts over two hundred gigawatts of data center capacity—roughly triple current levels. Annual electricity consumption by data centers approaches ten percent of national generation.[61,237]

The geographic distribution has shifted. Traditional hubs have reached capacity constraints. Growth has moved to secondary markets: Pennsylvania, Kansas, New Mexico, Mississippi. Rural America hosts gigawatt campuses where farmland stood five years earlier.

The power grid has transformed to serve this load. Natural gas plants built for data center demand account for a substantial fraction of new generation. Renewable capacity has expanded dramatically but falls short of covering data center growth. Nuclear plants scheduled for retirement remain online, and new small modular reactors are under construction—though none yet operational.

Residential electricity rates have risen fifteen to twenty-five percent in regions with heavy data center concentration, as utilities recover infrastructure investments. Capacity market prices have increased severalfold. In some regions, customers see higher charges without understanding the connection to facilities they have never seen.

Grid reliability has become contentious. Summer peaks strain systems designed for different load profiles. Rolling brownouts have occurred in Texas during extreme heat. The PJM grid serving the Mid-Atlantic issues emergency alerts with increasing frequency. Grid operators face impossible choices: curtail data center loads and risk contractual disputes, or curtail residential service and face political backlash. Most choose to build more generation, passing costs forward.

Carbon emissions from the electricity sector have not declined as planned. Data center load growth has offset efficiency gains and renewable additions. Climate advocates point to data centers as a major impediment to decarbonization. Operators counter with their renewable energy purchases, arguing that total emissions would be higher if AI workloads ran on less efficient infrastructure abroad.

Water consumption has become a constraint in arid regions. Arizona, Nevada, and parts of Texas have imposed restrictions effectively banning new evaporative-cooled facilities. Water-abundant regions attract proportionally more investment.

In Arizona, agricultural water allocations were cut repeatedly to accommodate industrial users. Farmers watched their allocations shrink by twenty, then thirty, then forty percent. Fields went fallow. The water that once grew food now cools servers.

———

In the exponential growth scenario, Saline Township in 2030 is a success story—at least by economic measures.

The original facility operates at full capacity, its 1.4 gigawatts of power flowing continuously to Oracle's cloud infrastructure. The equipment has been upgraded twice; the chips installed in 2026 were replaced in 2028 and again in late 2029. Each upgrade increased computational density while keeping power consumption roughly constant.

Employment has exceeded projections. The facility directly employs over six hundred people, more than the four hundred fifty originally promised. Wages have risen as operators compete for qualified technicians. The tight labor market has benefited existing residents who gained technical certifications.

Related Digital has expanded. A second campus, announced in 2028, rises on adjacent land. It will add another gigawatt of capacity, bringing total township power consumption above two gigawatts. DTE Energy has upgraded transmission infrastructure to serve the expanded load.

Tax revenue has followed the pattern established in the original project: meaningful but below projections, offset by state exemptions and assessment disputes. The township has used revenue gains to upgrade roads, expand the fire department, and establish a technology education program at the local school.

Not everyone is satisfied. Property values adjacent to the facilities declined and have not recovered. Traffic on Michigan Avenue has thickened. The rural character that attracted some residents has changed permanently. A small but vocal group continues opposing expansion, though they lack the numbers to affect policy.

The broader Ann Arbor region has felt spillover effects. Computer science enrollment at the University of Michigan has doubled since 2025. Technology companies have established regional offices. Housing prices have climbed, creating affordability challenges for residents not connected to the tech economy.

The exponential growth scenario produces economic gains but creates accumulating tensions.

Data center load growth strains grid planning. Interconnection queues remain long despite reforms. Some regions experience reliability events—not blackouts but brownouts and curtailments—when extreme weather coincides with peak demand. Ratepayers in some states see electricity bills rise significantly as utilities pass through infrastructure costs.

Communities hosting data centers divide between beneficiaries and those bearing costs. Political opposition organizes across regions, producing local moratoriums and stricter state regulations. The backlash does not halt development but redirects it toward more accommodating jurisdictions.

Climate targets become unachievable as data center emissions grow faster than reductions elsewhere in the economy. The gap between corporate sustainability claims and actual emissions triggers regulatory crises and public relations disasters for major operators.

Countries that did not invest as heavily in AI infrastructure fall behind in capabilities. Geopolitical tensions mount as the United States gains advantages in economic productivity, military applications, and scientific research. Some countries respond with protectionist policies; others attempt to attract investment with incentives surpassing what American states offer.

Labor market disruption extends far beyond data center operations. The infrastructure produces capabilities that reshape employment across the economy. Translation, tax preparation, technical writing, data entry, software development: the workflows rebuilt around AI inference displace workers by the millions. Entry-level positions suffer disproportionately, as Chapter 1 documented. By 2030, the World Economic Forum projects that 92 million jobs will be displaced even as 170 million new roles emerge.[238] But the displaced and the hired are not the same people. Net job creation means little to the fifty-year-old whose skills no longer command a salary.

The political implications are profound. White-collar work has been the primary conduit to middle-class status in the twenty-first century. College graduates expected stable careers in offices, not factories. When AI capabilities threaten those careers, the political response differs from manufacturing automation. Displaced factory workers organized unions and voted for protectionist trade policies. Displaced office workers— many of them in metropolitan areas, many with college degrees, many who supported technology as progress—face a different kind of reckoning. The infrastructure they helped finance through their retirement portfolios now threatens their employment.

By 2030 in this scenario, labor market anxiety has become a significant political force. Candidates campaign on AI regulation, job guarantees, and retraining programs. Some states impose taxes on automation; others subsidize human employment. The debate echoes the fights over trade policy that reshaped American politics in the 2010s—but the adversary is not China or Mexico. It is software running on silicon powered by electricity drawn from the grid, much of it generated in facilities like the one in Saline Township.

The irony compounds. Communities hosting data centers create hundreds of well-paid technician jobs. The same facilities enable AI systems that displace thousands of white-collar workers elsewhere. Rural America gains; suburban and urban America loses. The geography of winners and losers inverts the pattern of manufacturing automation, when rural factories closed while metropolitan economies thrived. This time the metropolitan knowledge workers feel the threat.

13.3   PLATEAU AND STRANDED ASSETS

In this scenario, AI demand growth slows dramatically after 2027, leaving the industry with more capacity than it can use profitably.

The scenario assumes AI capabilities plateau as models approach the limits of current architectures. Each new release offers incremental rather than transformational improvement. Enterprise adoption disappoints as companies struggle to deploy AI profitably and cut back on experimental budgets. Consumer applications saturate as the novelty of AI assistants wears off without compelling new use cases emerging. Capital flows away from data centers toward other sectors, with private equity firms pivoting to logistics and energy infrastructure. Completed projects struggle to find tenants. Construction halts on facilities that cannot secure anchor customers.

We have seen this pattern before. Fiber optics in the early 2000s. Cryptocurrencies in 2018 and 2022. Countless startups that raised money but never achieved sustainable businesses. This scenario assumes current enthusiasm has outrun sustainable demand.

Under plateau and stranded assets, the United States in 2030 has roughly one hundred gigawatts of data center capacity—significant growth from 2025 but well below exponential scenario levels. More troubling, utilization rates have declined. Many facilities operate at fifty to sixty percent capacity.

Excess capacity concentrates unevenly. Established markets maintain reasonable utilization. Rural campuses built on speculative demand struggle. Some projects stand only partially completed, construction halted when financing dried up.

The power grid has not transformed as expected. Planned generation projects were canceled when demand forecasts proved wrong. Transmission investments slowed. The anticipated strain on grid infrastructure never materialized because the loads never arrived.

Several large operators face financial distress. Private equity firms write down investments. The industry consolidates as weaker players sell to stronger ones.

Employment declines from its peak. Construction workers have moved on. Operating staff face layoffs as operators cut costs. Communities that expected permanent employment find the jobs more temporary than promised.

The carbon impact is lower than the exponential scenario but not as low as hoped. The renewable buildout scaled back along with investment.

In the plateau scenario, Saline Township's data center in 2030 operates but has not become the regional anchor boosters predicted.

The original facility reached full operation but has not expanded. The second campus was never built, with Related Digital citing "market conditions." Land assembled for expansion remains vacant.

Employment is below projections. Oracle reduced staffing as automation improved. The facility employs around three hundred fifty people—meaningful but fewer than promised.

Tax revenue is lower than expected. The facility's assessed value has been revised downward. Township officials face difficult choices.

Property values near the facility have not recovered. Rural character was lost, but the urban amenities that might have compensated never arrived.

The University of Michigan's expansion has moderated. Students who expected high-paying local jobs find fewer opportunities than anticipated.

Community opinion is divided. Some believe the project was a reasonable bet that did not pay off. Others believe they were deceived by projections that were never realistic.

———

The plateau scenario carries lessons beyond individual communities.

Facilities built at peak enthusiasm lose value as market conditions change. The question of who bears these losses—investors, communities, ratepayers—becomes politically contentious.

Infrastructure built to serve expected load has fewer customers than planned. Utilities must recover costs from a smaller base, raising rates. The alternative—writing off billions—harms shareholders and threatens utility stability. Either way, someone pays for capacity that sits unused.

States face criticism for giveaways that produced fewer benefits than promised. The experience makes them warier of future technology pitches.

The plateau has an unexpected beneficiary: workers whose jobs seemed threatened. AI capabilities that seemed poised to automate bookkeeping, market research, and HR screening prove less capable than feared—or prove capable but not cost-effective at scale. The labor displacement that seemed inevitable in 2025 materializes only partially. Office workers who spent years anxious about automation find their jobs persist, changed but not eliminated. The massive retraining programs proposed in the exponential scenario never prove necessary. This reprieve, however, comes with its own cost: the productivity gains that might have justified the infrastructure investment also fail to materialize. The economy neither transforms nor collapses. It muddles through.

The global AI industry consolidates around fewer, larger players.

13.4 TECHNOLOGY BREAKTHROUGH

In this scenario, a fundamental technology shift changes what AI infrastructure requires. The shift could take multiple forms. We explore one that seems plausible: a dramatic improvement in model efficiency that reduces computation requirements by orders of magnitude.

Imagine a breakthrough in neural network architecture that delivers efficiency gains comparable to what DeepSeek demonstrated in early 2025, but extended further. Models matching current frontier capabilities require one-tenth or one-hundredth the computation, collapsing the economics of hyperscale infrastructure. New training approaches eliminate the need for massive GPU clusters, allowing smaller labs to produce competitive models. Inference workloads migrate to edge devices—smartphones, laptops, local servers—rather than centralized data centers. Existing gigawatt-scale infrastructure becomes partially obsolete, valuable for some workloads but overbuilt for the industry's actual needs.

We have seen transitions like this before in computing. The shift from mainframes to minicomputers to personal computers disrupted each generation's infrastructure. The shift from on-premises servers to cloud computing stranded corporate data centers. Technology discontinuities reshape industries in ways difficult to predict in advance.

Consider the magnitude of previous transitions. The IBM System/360 Model 91, a 1966 scientific mainframe, consumed 74 kilowatts and delivered roughly two million floating-point operations per second. An iPhone 16 Pro fits in your palm, draws five watts, and delivers 2.5 trillion floating-point operations per second—over a million times the performance at one fifteen-thousandth the power. If AI follows a similar trajectory, the gigawatt training clusters of 2025 might seem as quaint as mainframe rooms by 2045. The implications for power demand would be profound: computing that today requires dedicated substations might eventually run on batteries. The trillion-dollar buildout assumes current architectures persist. History suggests otherwise.

———

For scenario planning purposes, we need not specify the breakthrough exactly. Consider one possibility: a shift from training massive models on enormous datasets to approaches that learn more efficiently from less data. Alternative techniques might achieve comparable capability with far less computation, fundamentally shifting infrastructure economics.

This scenario is not purely hypothetical. DeepSeek's January 2025 release shocked the AI industry by demonstrating that frontier performance does not require hyperscale infrastructure.[37,38] The Chinese lab claimed training costs of roughly six million dollars for a model matching capabilities that American labs spent hundreds of millions to achieve. The techniques were not secret—sparse architectures, lower-precision arithmetic, aggressive optimization—but DeepSeek applied them with unusual rigor.

The claim warrants scrutiny. It excludes research costs, failed experiments, and hardware acquisition. But even accounting for omissions, DeepSeek achieved frontier performance with dramatically less compute. NVIDIA lost $589 billion in market cap in a single day after the announcement, as investors priced in the possibility that the trillion-dollar buildout might prove unnecessary.[40]

Say, for argument's sake, that efficiency gains of ten-fold become standard across the industry by 2028. Models matching current frontier capabilities require one-tenth the compute. The gigawatt clusters built at great expense can be replaced by facilities orders of magnitude smaller.

Inference follows a similar pattern. Current AI inference happens in large data centers because models are too big and computationally intensive to run locally. A more efficient model might run on a smartphone, a laptop, or a small edge device. Centralized infrastructure that today seems essential becomes less necessary.

———

Under the technology breakthrough scenario, the United States in 2030 has AI capabilities exceeding the exponential scenario but far less infrastructure.

Total data center capacity is perhaps fifty gigawatts—significant growth from 2023 but dramatically less than exponential growth would have produced. Training happens in large facilities but requires far fewer resources. Inference has distributed to edge locations and personal devices.

Hyperscale campuses remain important but less dominant. Rural campuses built for massive training clusters have downsized or repurposed.

The power grid absorbed data center load without the strain that seemed inevitable in 2025. Load grew initially but plateaued in 2028 as efficiency gains outpaced demand.

Carbon emissions from AI are dramatically lower than any 2025 projection anticipated. The efficiency breakthrough reduced energy consumption per unit of output by ninety percent or more.

The technology industry has restructured. Companies that bet heavily on infrastructure find themselves with assets they do not need. Value has shifted from hardware and facilities to algorithms and applications.

In the breakthrough scenario, Saline Township's data center in 2030 exists but operates differently than planned.

Oracle reconfigured the facility after the efficiency breakthrough changed requirements. The 1.4-gigawatt campus runs at roughly half capacity. Some buildings are partially powered down; others converted to storage.

Employment is lower than any original projection. The facility employs around two hundred people. The technical staff who remain command high wages, but fewer are needed.

Tax revenue has declined along with operational intensity. The township faces budget adjustments.

The second campus was never built. Assembled land returned to agricultural use.

The regional economy adapted differently than expected. The Ann Arbor region became part of a distributed AI economy. The University of Michigan's computer science programs remain strong, but graduates work in software rather than infrastructure operations.

Community reaction is mixed. Opponents feel vindicated: the massive facility never fully materialized. Supporters feel the community benefited from infrastructure investment, even if outcomes differed from predictions. Whether that constitutes success depends on the baseline for comparison.

The breakthrough scenario creates clear winners and losers.

AI users gain powerful capabilities at a fraction of the cost. Communities that resisted data center proposals feel vindicated rather than left behind. The environment benefits from dramatically lower energy consumption.

The losers include everyone who bet on the old model. Data center developers find themselves with infrastructure that became unnecessary. Communities that offered generous incentives receive far less than projected. Workers who retrained for specialized facility operations find demand evaporating.

The breakthrough scenario illustrates a general principle: technology transitions create opportunities and disruptions alike. Those positioned for the previous era suffer; those positioned for the new era thrive.

13.5 COMPARING SCENARIOS

Comparing the scenarios clarifies how different futures affect the same community. Under exponential growth, the facility employs over six hundred people and expands to 2.4 gigawatts, with substantial regional spillover. Under plateau conditions, employment drops to around three hundred fifty, capacity remains at 1.4 gigawatts, with modest spillover. The breakthrough scenario produces the lightest footprint: approximately two hundred employees, capacity reduced to 0.7 gigawatts, with spillover taking a different character—distributed AI applications rather than infrastructure operations.

None of the scenarios produces the "quadruple tax revenue" originally promised, but all produce some economic benefit. The variation illustrates the uncertainty communities face.

***

The scenarios also differ in community character impact.

In the exponential scenario, Saline Township becomes industrialized. Multiple facilities span hundreds of acres. Traffic, noise, and visual impact are substantial. Rural character is definitively lost. Residents who valued that character adapt, relocate, or organize politically without much success.

In the plateau scenario, the community remains between worlds. One large facility exists but does not anchor broader development. Rural character is damaged but not fully replaced by industrial or suburban alternatives. The result is neither one thing nor the other.

In the breakthrough scenario, the community experiences a lighter touch. The facility exists and operates but at reduced intensity. Some land intended for data centers returns to other uses. The community is changed from 2025, but less changed than either alternative scenario projected.

The scenario comparison offers lessons for communities facing data center proposals.

Predictions are uncertain. Promoters who promise transformative benefits cannot know whether those benefits will arrive. Honest uncertainty should inform all projections.

Flexibility matters. Consent agreements that include provisions for multiple scenarios protect against futures that diverge from projections.

The baseline matters too. Evaluating a proposal requires comparing it to alternative development paths. What else might the land be used for? But communities should also ask whether the proposal is the only model available. The Deep Green model cannot serve hyperscalers, but it suggests alternatives exist. Communities facing gigawatt-scale proposals might reasonably ask why smaller, better-integrated facilities could not serve some of the same demand.

Values differ. Some prioritize jobs and tax revenue; others prioritize rural character and environmental quality. The challenge is building processes that acknowledge these differences.

13.6 LEADING INDICATORS

Which scenario will materialize? No one knows. But we can identify indicators that suggest movement toward one future or another.

Signs of exponential growth would include each new model release demonstrating substantial capability improvements over the current generation—GPT-5, Claude 5, Gemini 3. Enterprise customers would report measurable productivity gains and expand their AI budgets quarter after quarter. Consumer AI products would achieve daily active usage comparable to social media, with retention rates suggesting habit formation rather than novelty. Equinix, Digital Realty, and other operators would announce quarterly earnings that beat projections, with utilization above ninety percent. Electricity prices would remain stable despite growing data center demand, suggesting grid expansion is keeping pace. Generation and transmission projects would proceed on schedule, with interconnection queues clearing rather than lengthening. If we see these patterns, current growth trajectories are likely to continue.

As of January 2026, several indicators point toward exponential growth. OpenAI's Stargate initiative claims to be "well beyond halfway" to its 10-gigawatt goal in planned capacity after just one year.[207] Major hyperscalers continue announcing record capital expenditure budgets. Utilities report interconnection queues growing faster than they can process applications. These are early signals, not definitive proof, but they suggest the industry is not yet experiencing the demand plateau that would validate caution.

Signs of plateau would look different. New model releases would offer marginal improvements that reviewers describe as "incremental" rather than "breakthrough." Enterprise customers would publicly question AI ROI, with chief technology officers reporting that pilots did not scale. Consumer AI usage would plateau or decline, with apps dropping off home screens and subscriptions lapsing. Operators would report utilization rates below seventy percent, with facilities struggling to sign tenants at projected rates. Blackstone, KKR, and other private equity firms would redirect capital from data centers to logistics, energy, or other sectors. Construction projects would pause mid-build, with developers citing "market conditions" rather than specific financing failures. These patterns would suggest enthusiasm has outrun sustainable demand.

Signs of technological breakthrough would be different still. Peer-reviewed papers would demonstrate ten-times or hundred-times efficiency gains, with results reproducible by independent labs. Startups would train frontier-competitive models on budgets below ten million dollars, rather than the hundreds of millions current leaders spend. Apple, Qualcomm, and device manufacturers would announce on-device AI that matches cloud performance for common tasks. Hyperscalers would pause or cancel announced expansions, citing "reassessment of capacity requirements." AI capabilities would continue improving on benchmarks while total industry power consumption flattens or declines. Industry conferences and investor calls would focus on algorithm efficiency rather than infrastructure constraints as the key variable. These patterns would suggest technology is shifting in ways that invalidate assumptions underlying current infrastructure investment.

13.7 DECISION POINTS

The scenario analysis carries implications for actors making decisions today. Developers betting billions face substantial risk; the smartest protect themselves by preserving options rather than betting everything on one future. Policymakers offering incentives are betting with public money and should demand accountability: clawback provisions, transparency about project economics, coordination with neighboring jurisdictions to prevent bidding wars.

Communities evaluating proposals must decide under uncertainty. The safest assumption is that projections will be wrong in ways no one can predict. Communities should negotiate protections that work across scenarios and understand what is lost regardless of which scenario materializes: land, once paved, does not return to farmland; rural character, once industrialized, does not restore itself. The central question remains: what kind of community do residents want to live in?

Individual citizens watching the buildout may feel powerless. They are not. The citizens who shaped Saline Township's decision were people who cared about their community and showed up. That participation matters regardless of which scenario unfolds.

13.8 Wild Cards

Beyond the three main scenarios, certain developments could reshape outcomes dramatically regardless of baseline trajectory.

A major data center incident—environmental disaster, community health crisis, grid reliability failure—could trigger rapid regulatory change. Public opinion that currently tolerates expansion might shift overnight. A regulatory shock could pause development for years.

The buildout depends on continued capital flow. A broader financial crisis could interrupt that flow regardless of underlying demand for AI services. If credit tightens or recession reduces risk appetite, capital could evaporate. Projects halt, facilities change ownership, operators consolidate. Industry structure could transform even if AI demand remains strong.

Grid emergencies—severe storms, extended heat waves, cascading failures—could reveal vulnerabilities that normal operations conceal. If data centers are perceived as contributing to grid stress affecting residential customers, political tolerance for expansion might decline sharply.

The concentration of AI capability in a handful of facilities creates a different vulnerability. In April 2024, Ukrainian hackers destroyed a Russian datacenter, disrupting aerospace, telecommunications, and military operations.[239] The attack demonstrated that data centers are not merely commercial real estate but strategic targets. American facilities—hosting both civilian services and, increasingly, military AI applications—present similar opportunities to adversaries. A coordinated attack during a crisis could degrade commercial services and military capabilities simultaneously. Some operators have begun designing for resilience: behind-the-meter generation enables islanding during grid disruption; geographic distribution reduces concentration risk. Whether these measures would withstand state-level adversaries remains untested.

International events could reshape domestic priorities. A Taiwan Strait crisis threatening chip supply chains would create urgent pressure to expand domestic capacity. Conversely, continued policy shifts toward revenue-sharing rather than restriction could reduce the urgency framing that has accelerated approvals. The December 2025 H200 decision suggests that export policy will oscillate with administrations rather than move steadily toward decoupling. If Chinese customers can legally purchase powerful chips through American channels, the strategic justification for domestic infrastructure buildout shifts from national security toward commercial competition. De-escalation in US-China tensions could weaken expedited approvals further. Geopolitical disruption could accelerate or decelerate infrastructure development depending on its nature and the political response.

A wilder card still: what if the systems themselves become the variable? Some researchers assign meaningful probability to AI systems achieving capabilities that would render all other planning moot—not this decade, perhaps, but within the operational lifetime of facilities being built today. The scenarios in this chapter assume AI remains a tool. If that assumption fails, the infrastructure questions become either irrelevant or existential, depending on how the transition unfolds. Most industry participants dismiss such concerns as science fiction. A non-trivial minority lose sleep over them. The infrastructure gets built either way.

13.9 SCENARIO INTERACTIONS

The three main scenarios are not mutually exclusive. Regions might experience different scenarios simultaneously. Exponential growth could continue in some markets while plateau conditions emerge in others.

Markets with established customer bases remain attractive regardless of national trends. Markets built on speculation are more vulnerable. Stranded asset risk concentrates in some locations more than others.

Saline Township's anchor tenancy from Oracle provides some protection. But regional spillover effects depend on trends beyond the single facility.

Different scenarios might dominate at different times. Exponential growth through 2027 could give way to plateau conditions. A technology breakthrough in 2028 could reshape requirements regardless of what came before. Timing uncertainty compounds trajectory uncertainty.

The scenarios interact through feedback effects. Plateau conditions might trigger consolidation, restoring utilization and enabling renewed growth. Exponential growth straining the grid could trigger regulatory responses that slow development. These feedback effects dampen extreme outcomes. Reality will oscillate rather than follow a smooth path.

The three scenarios examined represent dramatic outcomes: exponential transformation, painful contraction, technological disruption. But the most likely future may be moderate growth that neither transforms nor devastates.

Under moderate growth, AI demand continues expanding at rates closer to fifteen or twenty percent annually rather than fifty or one hundred. Some projects succeed; others struggle. The industry consolidates around sustainable economics.

This scenario receives less attention because it lacks drama. The boring middle ground does not motivate investment pitches or policy urgency.

The telecom bubble of the late 1990s offers perspective. Companies invested more than 500 billion dollars in fiber optic infrastructure, claiming internet traffic was "doubling every 100 days." Traffic actually doubled roughly once per year—a fourfold overestimate that compounded annually.[179] By 2002, ninety-five percent of installed fiber remained dark. Twenty-three telecom companies went bankrupt. Five hundred thousand workers lost jobs.[180]

But the fiber eventually found use. YouTube launched in 2005. Netflix streaming followed in 2007. Cloud computing expanded through the 2010s. Infrastructure built for exaggerated demand eventually served real demand—but years later than investors expected, after enormous losses.

The data center buildout could follow a similar pattern. Demand projections prove optimistic. Investment exceeds near-term requirements.

Consolidation follows. Eventually the infrastructure serves purposes not fully anticipated.

For communities like Saline Township, moderate growth might be the most ambiguous outcome. The facility operates but at reduced intensity. Jobs exist but fewer than promised. The community is neither transformed nor abandoned—merely changed in ways that resist simple evaluation.

13.10 THE CERTAINTY OF UNCERTAINTY

Reality will not match any scenario precisely. The actual future will combine elements of multiple scenarios, unfold differently in different regions, and include developments no scenario anticipated.

Anyone claiming to know how AI infrastructure will evolve through 2030 is either deceiving others or deceiving themselves. The uncertainty extends not just to how much infrastructure gets built, but to what kind—whether the greenfield gigawatt model continues unchallenged or whether alternatives gain ground.

This uncertainty does not mean decisions cannot be made. It means decisions should acknowledge uncertainty: preserve flexibility where possible, protect against downside risks where necessary, and make choices based on values that persist across scenarios. The same uncertainty that makes predictions unreliable also means the future remains open to different choices—about where to build, at what scale, and who benefits.

13.11 PREPARING FOR MULTIPLE FUTURES

Given uncertainty, how should actors prepare? The most valuable position is one that preserves options.

For investors, optionality means avoiding overleveraged positions. Diverse portfolios across regions, facility types, and tenants reduce dependence on any single trajectory.

For operators, optionality means designing facilities that can serve multiple purposes. Buildings that can only serve AI training workloads are more vulnerable than buildings that can adapt.

For communities, optionality means agreements that allow adjustment. Review points, performance triggers, renegotiation provisions preserve options. But it also means demanding that policymakers create options worth preserving—incentive structures that reward integration, not just speed; permitting processes that make brownfield viable, not just legal.

Beyond preserving options, actors can explicitly hedge against adverse outcomes. Investors can diversify. Operators can contract with multiple customers. Communities can diversify economic development—and can advocate for policies that bring development to places that need it rather than places that happen to have transmission lines crossing cornfields.

The Saline Township consent agreement includes some hedging elements: employment and infrastructure provisions provide benefits beyond tax revenue. More comprehensive hedging would provide stronger protection. Better policy would make such hedging unnecessary by aligning developer incentives with community interests from the start.

Under uncertainty, rapid learning has value. Actors who recognize early whether trends match expectations can adjust faster.

For the industry, this means monitoring indicators: AI demand metrics, utilization rates, capital flows. For communities, it means tracking whether promised benefits materialize. For citizens, it means watching whether policy reforms gain traction—and pushing harder if they do not.

Saline Township decided under uncertainty, within constraints it did not create. The incentive structure, the timeline pressure, the legal asymmetry—all of these shaped a decision space that made greenfield development nearly inevitable. Other communities will face similar pressures. Some will follow the same path. A few may find room for different choices, if policy creates that room.

The scenarios explored here provide a framework for thinking through possibilities. They do not tell communities what to decide. They help communities understand what they are deciding about—and what alternatives might exist if they demand them.

13.12 A Different Model

The three scenarios explored in this chapter share an assumption worth examining. All three presume that AI infrastructure continues following the current model: gigawatt-scale facilities on greenfield agricultural land, built by hyperscalers and private equity, connected to regional grids through multi-year interconnection processes. The scenarios vary in how much of this infrastructure gets built. They do not question whether this model is the only one available.

It is not.

————

Sixty-five miles northwest of Saline Township, Lansing faces a different problem. The city's downtown district heating system, a network of underground steam pipes that warms government buildings, hospitals, and commercial properties, is over a century old and failing. City officials have watched the maintenance bills climb for years. Replacing the system would cost tens of millions of dollars the city does not have. Abandoning it would force dozens of buildings to install individual heating systems, a distributed expense that would fall on property owners already struggling with downtown vacancy.

Deep Green, a UK-based company, proposed a solution: a 24-megawatt data center on an unused parking lot in the stadium district whose waste heat would feed the city's heating network.[66] The facility would replace fossil fuel heating with recovered thermal energy. The city would gain a revenue stream and solve its infrastructure problem. Deep Green would gain a customer for heat that other data centers simply exhaust to the atmosphere. No farmland would be converted. No township would be transformed.

The model works because it matches facility scale to urban infrastructure needs. A 24-megawatt facility produces heat at volumes a city heating network can absorb. A 1.4-gigawatt facility like Saline's would overwhelm any urban system—and would never fit in an urban footprint regardless. Deep Green is not competing with Oracle for the same workloads. It serves different customers: universities, engineering firms, CGI studios, enterprises that need computing but not at hyperscale.

The Lansing City Council is scheduled to vote on the project in February 2026. The outcome is uncertain. But the model has been demonstrated: a data center that solves a city's problem rather than creating a township's dilemma.

———

Deep Green cannot replace Saline. The workloads are different; the scale is different; the customers are different. But the question is not whether one model can substitute for another. The question is whether policy should encourage both models or only one.

Current policy overwhelmingly favors the greenfield gigawatt model. State incentives apply equally regardless of site type: sales tax exemptions, property tax abatements, expedited permitting. When a developer can build faster and cheaper on farmland, equal incentives mean farmland wins. The timeline penalty for brownfield sites, two to three years of additional remediation and permitting, receives no compensation. The community benefit of urban integration receives no reward—solving infrastructure problems, revitalizing blighted areas, creating jobs where people already live.

Michigan attempted to change this in January 2025, offering tax exemption through 2065 for brownfield sites versus 2050 for greenfield.[80] The fifteen-year differential aimed to offset brownfield timeline costs with extended benefits. The logic was sound. The politics were not: by December 2025, bipartisan legislation sought to repeal the program before it could demonstrate results. Greenfield projects captured most benefits while brownfield uptake remained minimal. The experiment ended before it began.

But the failure was one of implementation, not concept. Incentives scaled to facility size would encourage the Deep Green model without the fiscal exposure that doomed Michigan's approach. Stronger benefits for smaller urban-integrated projects would shift the calculus. A 24-megawatt facility receiving a 50 percent property tax abatement costs a city far less than a 1.4-gigawatt facility receiving the same percentage. Fast-track permitting for brownfield sites, offsetting the timeline penalty that currently makes them uncompetitive, would require regulatory coordination but no public expenditure. Utility support for urban interconnection would redirect infrastructure investment rather than increasing it—prioritizing grid upgrades that enable brownfield development instead of rural greenfield expansion.

The Inflation Reduction Act's Energy Community Bonus points in the right direction: a ten percent tax credit bonus for clean energy projects on brownfield or coal community sites.[134] But this bonus applies to power generation, not consumption. A data center on a brownfield site receives no federal preference over one on a cornfield. Extending similar logic to large electricity consumers would align incentives with stated policy goals.

Lansing is not unique. Dozens of American cities share its characteristics: aging infrastructure that data centers could help modernize, brownfield sites awaiting redevelopment, populations that would benefit from technical employment, political will to try something different. The geography of need does not match the geography of investment.

Detroit has more brownfield acreage than any city in America: former auto plants, steel mills, and manufacturing facilities sitting idle while data centers rise on farmland an hour away. Gary, Indiana, once produced more steel than any city on Earth; now it has vacant land, available power, and a population that has declined by two-thirds since 1960. Youngstown, Birmingham, St. Louis—the list extends across the Rust Belt and into the Sun Belt, cities that lost manufacturing jobs decades ago and have not found replacements.

These cities cannot host gigawatt-scale hyperscale facilities. The infrastructure gaps are too large, the remediation too complex, the political negotiations too slow. But they could host Deep Green-scale facilities: 20 to 50 megawatts, serving regional enterprise customers, integrated with urban systems rather than isolated on rural land.

The arithmetic is not trivial. Fifty facilities at 24 megawatts each total 1.2 gigawatts—roughly the capacity of the Saline project. The same computing capacity, distributed across fifty cities instead of concentrated in one township. The same jobs, located where people already live instead of where workers must relocate. The same tax revenue, flowing to municipalities that desperately need it instead of to townships that never asked for transformation.

This is not a fantasy. It is a policy choice. The current policy chooses differently.

The choices being made now will structure possibilities for decades. Incentive programs enacted in 2025 will govern projects through 2050.

Transmission infrastructure approved this year will determine which sites can receive power in 2030. Permitting precedents established in early cases will shape how regulators evaluate later ones.

Readers who care about farmland preservation have standing to demand that incentives distinguish between greenfield and brownfield sites. Readers who care about urban revitalization have standing to demand that policy reward facilities solving city infrastructure problems. Readers who care about equitable development have standing to ask why computing jobs concentrate in wealthy suburbs while struggling cities watch from the sidelines.

These are not technical questions with technical answers. They are political questions that will be resolved through political processes. State legislatures set incentive structures. Public utility commissions approve interconnection priorities. City councils vote on zoning and development agreements. Federal agencies write rules implementing laws that Congress passes.

The concrete is not yet poured on most of the facilities announced in this buildout. Press releases are not shovels. The decisions that will determine where concrete actually gets poured are being made now, in processes that remain open to public input. Farmland or brownfield. Townships or cities. Extraction or integration. These choices remain open.

Showing up matters. The residents of Saline Township who packed the board meeting in September 2025 did not stop the project. But they shaped the consent agreement. They established that communities have standing to negotiate. They demonstrated that showing up changes outcomes, even when it does not reverse them.

The same is true for the policy choices that determine which model prevails. State legislators considering incentive reforms will hear from developers who prefer the status quo. They will hear from industry associations defending current arrangements. Whether they also hear from constituents who want different priorities depends on whether those constituents show up—at hearings, in comment periods, through their representatives.

The future is not determined. It is being chosen. The scenarios in this chapter describe what might happen if current trajectories continue. This section describes what could happen if enough people demand a different trajectory.

The epilogue returns to Saline Township one final time. Not to judge the decision, but to witness what was chosen and what was lost.

The Token, Revisited

January 2030

FRANK left the township board in 2028. He cited his grandchildren when he stepped down—*Lily* is twelve now, *Marcus* ten, and they still ask him about the big computer—and a desire to spend winters somewhere warmer. But he has not moved yet. Something keeps him here. His wife says it is stubbornness. He thinks it might be something else. Witness, maybe. The need to see what he helped create.

His daughter *Beth* works in Troy now, in mortgage processing for a regional bank. She handles the files the AI flags as exceptions—the applications with irregular income, the self-employed borrowers whose tax returns do not fit the templates, the refinances where the appraisal comes in wrong. The algorithms process a thousand files for every one she touches. She is always on call. Her phone buzzes at dinner, at *Marcus*'s soccer games, at 2 a.m. when a closing is scheduled for the next morning and something has triggered a fraud alert. The software never sleeps, and someone has to approve the decisions it cannot make alone. Her salary has not risen in three years, though her hours have. She tells *Frank* that she is lucky to have the job at all. He notices she looks tired.

He thinks about that vote every day. Not whether it was right or wrong—he has made his peace with that. He thinks about whether they could have done it differently. Whether there was ever really a choice.

He keeps in touch with *Ellen*, whose family still farms the adjacent land. They still have coffee at City Limits Diner—the same booth, the same view of Michigan Avenue, the same waitresses who stopped asking what they wanted years ago. The diner sits across from Oakwood Cemetery, next to American Legion Post 322, a few hundred feet from Ford's old soybean plant. In the 1930s, Ford bought crops from seven hundred local farmers and processed them into oil for paint and plastics—part of his "Village Industries" program, where rural workers could draw factory wages without leaving their land. The plant closed after his death. The building became an antique shop, then an event venue. The dam still holds.

They sit in the same booth each time, the one by the window with the longest cathedral grains in the table. A framed aerial photo of old Saline hangs on the wall—the town before the subdivisions and the strip malls, when the farms ran unbroken to the horizon. The menu still runs from goulash to gluten-free. The coffee comes in heavy ceramic mugs that have outlasted most marriages. Semis carrying equipment pass on the same road that once carried Model Ts and covered wagons. They watch them go by, the way people in this booth have watched traffic go by for forty years.

Her well never ran dry. The consent agreement worked, at least for that. *Ellen* retired from teaching in 2027—thirty years at Saline High School, three decades of biology students who now work in hospitals and labs and, yes, data centers across the country. She can still identify every bird on her property by its song, though some songs she hears less often now. The cardinals stayed. The meadowlarks left when the fields did.

Her sons never developed any interest in farming. The older one works at the university, using artificial intelligence to find patterns in genomic data—she understood the genetics from thirty years of teaching, but not what the AI was doing with it. The younger one's company outgrew Ann Arbor; he moved to Detroit in 2027. His software reads contracts, flags issues, summarizes terms. He tried to explain it to her once, over Sunday dinner. She listened and nodded and thought about her former students who became paralegals. When the company raised its funding round, he was promoted. When AI tools started writing code as well as junior developers in 2029, his team shrank. He does not talk about work anymore when he visits. She does not ask.

Once, over Thanksgiving, after too much wine, he said something that stayed with her. "Mom, some of the people I work with—the serious ones, the ones who understand the math—they're not worried about losing their jobs. They're worried about losing everything. They have these probabilities they assign. P of doom, they call it. The chance that we build something we can't control." He laughed, but it wasn't a real laugh. "I don't know what number I'd put on it. But it's not zero. And I help build it anyway. Every day." She did not know what to say. She still doesn't.

The irony is not lost on her: her son builds tools that run on the same infrastructure she fought against, and those tools now threaten his own position—and perhaps, if the serious ones are right, threaten something larger still. Eventually the land will sell. She knows this now. She has made a kind of peace with it, the way you make peace with weather you cannot change. The philodendron her mother planted in 1978 still sits on her kitchen windowsill, in the same spot where her mother served breakfast for forty years. She waters it every Sunday morning, at the oak table scarred by a thousand family meals. Some things you can keep.

"She doesn't blame me," *Frank* says. "I asked her directly, a couple years ago. We were having coffee, watching the trucks go by, and I just asked. She said she understood. The township couldn't have won. The best we could do was get what we got." He pauses, turns his coffee cup in his hands, the honey-colored wood rail smooth against his elbow. The ceramic is warm. The coffee is cold. He drinks it anyway, sets it down—

the soft knock of ceramic meeting oak milled before he was born. "That's probably true. I hope it's true. I need it to be true."

Harold moved to Florida in 2027, to a condo near his daughter in Naples. The farm money set up his grandchildren for college, paid off debts he had carried for decades, bought him comfort he had never known. He died in 2029, quietly, in his sleep, with the air conditioning humming and palm trees outside his window instead of oaks. *Ellen* heard from his daughter. She called *Frank* that evening, and they sat together in the diner booth for an hour without saying much of anything, the booth backs' patchwork of dusty blues fading in the evening light. The waitress re-filled their coffee twice without being asked. *Harold* had never come to those Sunday dinners she kept inviting him to. Now he never would.

"He understood," *Ellen* said finally. Her voice was rough. She had been crying before she got there; her eyes were red, and she kept pressing a napkin to them. "What we were losing. Even if he took the money. Especially because he took the money."

Frank nodded. He thought about that phone call years ago, when *Harold* had asked what he should do, and *Frank* had said he understood either way. He still meant it. He hoped *Harold* had known that, at the end. Outside the window, a semi-truck rumbled past, headed toward the facility. They watched it go.

David left the private equity fund in 2029. The returns had been strong—his investors were satisfied, his partners were wealthy—but something had shifted. He tells people he wanted to spend more time with his daughters, which is true. His oldest is in high school now, the one who showed him ChatGPT back in 2023 and changed his career. He missed too many of her volleyball games. He does not want to miss more.

The export control whiplash contributed too, though he does not talk about it much. Three policy reversals in 2025 alone. The H200 approval that December changed competitive dynamics overnight. By 2027, some facilities his fund had financed faced Chinese competition they had not modeled. The deals still worked, mostly. But the policy uncertainty wore on him. Building infrastructure for decades when policy shifts every administration felt like betting on a roulette wheel. He grew tired of recalculating.

He does not tell people about the sleepless nights, about the spreadsheets that stopped making sense. Somewhere along the way, the numbers had detached from anything real. A billion dollars became a rounding error. A thousand acres became a cell in a spreadsheet. The land had names once, families who worked it for generations. By the end, it was just coordinates and zoning classifications.

He thinks about *Frank* sometimes, though they never met. He read about the Saline Township consent agreement in the trade press. He recognized the pattern: local resistance, legal pressure, settlement, capitulation. His fund had been on the other side of similar negotiations. They had won, mostly. He is no longer sure what winning means.

"I still believe in what we built," he says, when asked. "The infrastructure matters. AI matters. But we could have done it differently. We could have listened more. We did not have to treat every community like an obstacle." He pauses. "I am not sure we knew how."

The buildout is not over. Decisions made now will shape what comes next; precedents set now will guide projects for decades. The questions that *Frank* and *Ellen* faced will confront other communities, in other states, with other resources and other vulnerabilities.

The miracle has faded from view. The costs have not.

———

Frank drives past the facility one more time. His wife asked him to pick up groceries, but he took the long way, the way that passes the property line. He takes Michigan Avenue, the old US-12—the road they called the Chicago Pike when pioneers first cut it through the wilderness in 1827, following Indian trails west toward a city that barely existed. His truck knows the route by heart. The same F-150 he has had for eleven years, 187,000 miles on the odometer.

The red barn catches afternoon light, the same light that has fallen on this land for centuries, long before his grandfather was born, long before anyone thought to build a metropolis of silicon on a cornfield. He remembers when it was just a barn surrounded by fields, *Harold*'s barn. The barn where *Harold* taught him to swap a filter, to cut down a PTO. Before anyone had heard of tokens or gigawatts or Stargate. Before the vote. Before the lawsuit. Before.

But now the buildings hum, not the tractors. *Frank* remembers the sound of *Harold*'s Deere from half a mile off. A diesel throb you felt in your chest before it reached your ears. This hum is different. Higher. Steadier. It never stops.

He thinks about what his father taught him. How you put seed in the ground and then you wait. You check the weather. You walk the rows. You watch for blight, for bugs, for the late frost that kills everything. And none of it is in your hands. You did the work. You spent the money. Now you wait to see if the world lets you keep what you planted.

That is what this feels like now. The concrete is poured. The copper is run. The chips are spinning. Somewhere, men in offices are watching numbers the way *Harold* used to watch the sky. Calculating yields. Praying for rain they cannot make. Hoping the harvest comes in before something fails. Before the money runs out. Before the weather turns.

Frank's father farmed eighty acres until 1974. He remembers helping bring in the corn, the exhaustion of it, the relief when the trucks pulled away full. He remembers the year the rain didn't come. His father walking the dying rows every evening, touching the curled leaves like he could will them back to life. There was nothing to do. The seed was in the ground. The money was spent. All you could do was wait and hope.

381

Nobody talks about the waiting. The months between planting and harvest when every storm makes your chest tight, every clear morning feels like borrowed time. Farmers know. You work and you spend and you hand it over to forces you cannot control. Weather. Markets. Luck. You find out in September whether you were right or wrong.

He watches the facility hum. Somewhere in those buildings, silicon is doing whatever silicon does. The seeds are different now. But the waiting is the same.

Somewhere, someone types a question.

But it does not matter if another question comes. It does not matter if the money men get their harvest in. The soil is already under concrete. The farms are gone.

This is server country now.

Bibliography

[1] Brian Allnutt. "Data center developer takes Saline Township to court over rezoning decision". In: *Planet Detroit* (Sept. 18, 2025).

[2] OpenAI and SoftBank. *Announcing The Stargate Project*. OpenAI. Jan. 21, 2025.

[3] Saline Township Board. *Special Meeting Minutes, September 24, 2025*. Saline Township. Sept. 24, 2025.

[4] Saline Township Board. *Special Meeting Minutes, October 1, 2025*. Saline Township. Oct. 1, 2025.

[5] Washtenaw County Circuit Court. *Consent Judgment — RD Michigan Property Owner I LLC et al. v. Saline Township*. Saline Township. Oct. 15, 2025.

[6] Clare Duffy. "Trump announces a $500 billion AI infrastructure investment in the US". In: *CNN Business* (Jan. 21, 2025).

[7] Michigan Attorney General. *AG Nessel on the MPSC Approving DTE's Application to Service Saline Data Center*. State of Michigan. Dec. 18, 2025.

[8] Michael J. Bommarito. *U.S. Data Center Project Database*. 2025.

[9] Brad Smith. *The golden opportunity for American AI*. Microsoft. Jan. 3, 2025.

[10] Tom B. Brown et al. "Language Models are Few-Shot Learners". In: *Advances in Neural Information Processing Systems*. Vol. 33. 2020, pp. 1877–1901.

[11] Josh You. *How much energy does ChatGPT use?* Epoch AI. Feb. 7, 2025.

[12] Amin Vahdat and Jeff Dean. *How much energy does Google's AI use? We did the math*. Google Cloud. Aug. 21, 2025.

[13] Amanda Silberling. "ChatGPT users send 2.5 billion prompts a day". In: *TechCrunch* (July 21, 2025).

[14] David Cahn. *AI's $600B Question*. Sequoia Capital. June 20, 2024.

[15] Microsoft Corporation. *Earnings Release FY25 Q3: Microsoft Cloud and AI Strength Drives Third Quarter Results*. Microsoft Corporation. Apr. 30, 2025.

[16] NVIDIA Corporation. *NVIDIA H100 Tensor Core GPU Datasheet*. NVIDIA Corporation. 2022.

[17] International Energy Agency. *Energy and AI*. International Energy Agency, Apr. 2025.

[18] Annalise Frank. "OpenAI data center planned near Ann Arbor". In: *Axios Detroit* (Oct. 31, 2025).

[19] Erik Brynjolfsson, Bharat Chandar, and Ruyu Chen. *Canaries in the Coal Mine? Six Facts about the Recent Employment Effects of Artificial Intelligence*. Tech. rep. Stanford Digital Economy Lab, Nov. 2025.

[20] Mark Muro, Shriya Methkupally, and Molly Kinder. "The Geography of Generative AI's Workforce Impacts Will Likely Differ from Those of Previous Technologies". In: *Brookings Metro* (Feb. 2025).

[21] Microsoft Corporation. *Microsoft will be carbon negative by 2030*. Microsoft Corporation. Jan. 16, 2020.

[22] Google. *Our third decade of climate action: Realizing a carbon-free future*. Google. Sept. 14, 2020.

[23] Kif Leswing. "Nvidia dominates the AI chip market, but there's more competition than ever". In: *CNBC* (June 2, 2024).

[24] NVIDIA Corporation. *NVIDIA Blackwell Platform Arrives to Power a New Era of Computing*. NVIDIA Corporation. Mar. 18, 2024.

[25] NVIDIA Corporation. *NVIDIA GB300 NVL72 Specifications*. NVIDIA Corporation. 2025.

[26] Alex Krizhevsky, Ilya Sutskever, and Geoffrey E. Hinton. "ImageNet Classification with Deep Convolutional Neural Networks". In: *Advances in Neural Information Processing Systems*. Vol. 25. 2012.

[27] NVIDIA Corporation. *NVIDIA Announces Financial Results for Fourth Quarter and Fiscal 2024*. NVIDIA Corporation. Feb. 21, 2024.

[28] AMD. *AMD Instinct MI300X Accelerators*. AMD. Dec. 6, 2023.

[29] Datacenters.com Development. "Crusoe's $400M AMD Deal: A Game-Changer for AI Data Centers". In: *Datacenters.com* (July 7, 2025).

[30] Google Cloud. *Introducing Trillium, sixth-generation TPUs*. Google Cloud. May 14, 2024.

[31] Amazon Web Services. *AWS Trainium*. Amazon Web Services. 2024.

[32] Andrew Ling. *Inside the LPU: Deconstructing Groq's Speed*. Groq. Aug. 1, 2025.

[33] Kavout. "NVIDIA's $20 Billion Groq Deal: What It Means for AI

in 2026". In: *Kavout Market Lens* (Dec. 27, 2025).

[34] Anton Shilov. "Intel launches Gaudi 3 accelerator for AI: Slower than Nvidia's H100, but also cheaper". In: *Tom's Hardware* (Sept. 24, 2024).

[35] Anton Shilov. "Intel expects paltry $500 million in Gaudi 3 AI sales for the rest of the year". In: *Tom's Hardware* (Apr. 30, 2024).

[36] Duncan Stewart et al. *More compute for AI, not less.* Deloitte, Nov. 18, 2025.

[37] Kyle Wiggers. "DeepSeek claims its reasoning model beats OpenAI's o1 on certain benchmarks". In: *TechCrunch* (Jan. 27, 2025).

[38] DeepSeek AI. *DeepSeek-V3 Technical Report.* Tech. rep. DeepSeek, Dec. 27, 2024.

[39] Hayden Field. "DeepSeek's hardware spend could be as high as $500 million, new report estimates". In: *CNBC* (Jan. 31, 2025).

[40] Gregory C. Allen. "DeepSeek, Huawei, Export Controls, and the Future of the US-China AI Race". In: *Center for Strategic and International Studies* (Mar. 7, 2025).

[41] Jianxian Wu. "Digital Jevons paradox in urban data center energy systems". In: *Nature Cities* 2 (July 15, 2025), p. 677.

[42] Satya Nadella. *Jevons paradox strikes again!* X (formerly Twitter). Jan. 27, 2025.

[43] Bhargs Srivathsan et al. *AI Power: Expanding Data Center Capacity to Meet Growing Demand.* McKinsey & Company, Oct. 29, 2024.

[44] Chhavi Arora et al. *The Next Big Shifts in AI Workloads and Hyperscaler Strategies.* McKinsey & Company, Dec. 17, 2025.

[45] Michael Acton. "Intel shares slide 11% as supply issues limit growth". In: *Financial Times* (Jan. 22, 2026).

[46] The White House. *FACT SHEET: President Biden Announces Up to $8.5 Billion Preliminary Agreement with Intel under the CHIPS and Science Act.* The White House. Mar. 20, 2024.

[47] U.S. Department of Commerce, Bureau of Industry and Security. *Commerce Implements New Export Controls on Advanced Computing and Semiconductor Manufacturing Items to the People's Republic of China.* U.S. Department of Commerce. Oct. 7, 2022.

[48] Alex Haag. *The State of AI Competition in Advanced Economies.* Federal Reserve Board. Oct. 6, 2025.

[49] Kevin Lee and Shubho Sengupta. *Introducing the AI Research SuperCluster — Meta's cutting-edge AI supercomputer for AI research.* Meta AI. Jan. 24, 2022.

[50] NVIDIA Corporation. *NVIDIA GB200 NVL72 Rack-Scale System*. NVIDIA Corporation. 2025.

[51] CyrusOne. *CyrusOne Introduces Intelliscale – The Future of AI Workload Data Centers*. CyrusOne. Aug. 24, 2023.

[52] NVIDIA Corporation. *NVIDIA Ethernet Networking Accelerates World's Largest AI Supercomputer, Built by xAI*. NVIDIA Corporation. Oct. 28, 2024.

[53] CoreWeave, Inc. *Form S-1 Registration Statement*. U.S. Securities and Exchange Commission. Mar. 3, 2025.

[54] Dan Swinhoe. "Meta to invest "hundreds of billions of dollars into compute to build superintelligence," with several multi-GW data center clusters". In: *Data Center Dynamics* (July 14, 2025).

[55] Donald J. Trump. *Remarks on Artificial Intelligence Infrastructure Development and an Exchange With Reporters*. The American Presidency Project. Jan. 21, 2025.

[56] Brian Allnutt. *Saline Township settles with data center developer: 'Lesser of two evils'*. Planet Detroit. Oct. 21, 2025.

[57] Kyle Davidson. *Whitmer: Multi-billion-dollar Saline Township data center 'largest investment in Michigan history'*. Michigan Advance. Oct. 30, 2025.

[58] Related Digital. *Michigan — RELATED DIGITAL*. Related Digital. 2025.

[59] Vertiv. "Understanding direct-to-chip cooling in HPC infrastructure: A deep dive into liquid cooling". In: *Vertiv Technical Articles* (2024).

[60] Seamus Nayduch. *CoreWeave Data Center Operations: Built for AI*. CoreWeave. Feb. 18, 2025.

[61] Arman Shehabi et al. *2024 United States Data Center Energy Usage Report*. Lawrence Berkeley National Laboratory, Dec. 2024.

[62] Miguel Yañez-Barnuevo. "Data Centers and Water Consumption". In: *Environmental and Energy Study Institute* (June 25, 2025).

[63] Yana Kunichoff. "Tucson City Council rejects Project Blue data center amid intense community pressure". In: *AZ Luminaria* (Aug. 6, 2025).

[64] Applied Digital. *Applied Digital Achieves Ready for Service for Phase 1 at Polaris Forge 1 Building 1 for CoreWeave*. Applied Digital. Oct. 27, 2025.

[65] Kelly House and Paula Gardner. "Tech giants announce $7B data center, Michigan's first hyperscale campus". In: *Bridge Michigan* (Oct. 30, 2025).

[66] Leo V. Kaplan. "Deep Green data center: Where do things stand?" In: *Lansing City Pulse* (Jan. 21, 2026).

[67] Bloomberg. "British Airways Owner Says Data Center Outage Cost £80M". In: *Data Center Knowledge* (June 15, 2017).

[68] PJM Interconnection. *2025 PJM Long-Term Load Forecast Report.* PJM Interconnection, Jan. 2025.

[69] Lawrence Berkeley National Laboratory. *Queued Up: Characteristics of Power Plants Seeking Transmission Interconnection.* Lawrence Berkeley National Laboratory. 2025.

[70] ERCOT. *Large Load Integration.* Electric Reliability Council of Texas. 2024.

[71] PJM Interconnection. *2025/2026 Base Residual Auction Report.* PJM Interconnection, July 30, 2024.

[72] CBRE. *North America Data Center Trends H1 2025.* CBRE, Sept. 2025.

[73] CBRE. *2025 Global Data Center Investor Intentions Survey.* CBRE, 2025.

[74] Dominion Energy. *Q4 2024 Earnings Conference Call Slide Presentation.* Dominion Energy. Feb. 12, 2025.

[75] Alander Rocha. "Georgia regulators approve massive power grid expansion to serve data centers". In: *Georgia Recorder* (Dec. 19, 2025).

[76] Entergy Louisiana. *Entergy Louisiana receives LPSC approval for major infrastructure investments to support Meta's data center and*

improve reliability. Entergy Corporation. Aug. 20, 2025.

[77] David Harvey. *Spaces of Capital: Towards a Critical Geography.* New York: Routledge, 2001.

[78] Charlotte Trueman. "Nvidia data center GPU shipments totaled 3.76m in 2023, equating to a 98% market share - report". In: *Data Center Dynamics* (June 12, 2024).

[79] Microsoft Corporation. *Fourth Quarter Fiscal Year 2024 Earnings Conference Call.* Microsoft Corporation. July 30, 2024.

[80] Michigan Legislature. *Senate Bill 237 of 2023 (Public Act 181 of 2024).* Dec. 30, 2024.

[81] Brian Allnutt and Nina Misuraca Ignaczak. "Michigan Senate approves data center tax breaks that environmental advocates say will kill the state climate plan". In: *Planet Detroit* (Dec. 14, 2024).

[82] Mike Duffy. "Michigan Attorney General Dana Nessel questions redacted DTE data center contracts in Saline Township". In: *WXYZ Detroit* (Dec. 16, 2025).

[83] PJM Interconnection. *PJM at a Glance.* PJM Interconnection. 2025.

[84] Federal Energy Regulatory Commission and North American Electric Reliability Corporation. *The February 2021 Cold Weather Outages in Texas and the South Central United States.* Federal

Energy Regulatory Commission, Nov. 2021.

[85] Texas Department of State Health Services. *February 2021 Winter Storm-Related Deaths — Texas.* Texas Department of State Health Services, Dec. 31, 2021.

[86] Lana Ferguson. "Texas' data center boom contributes to ERCOT's large load requests quadrupling in 2025". In: *The Dallas Morning News* (Dec. 9, 2025).

[87] Zachary Skidmore. "VoltaGrid to supply Oracle with 2.3GW of natural gas power for AI data centers". In: *Data Center Dynamics* (Oct. 16, 2025).

[88] Amy Joi O'Donoghue. "Millard County data center to break ground and bring promise to rural Utah". In: *Deseret News* (Oct. 21, 2025).

[89] Tim De Chant. "xAI gets permits for 15 natural gas generators at Memphis data center". In: *TechCrunch* (July 3, 2025).

[90] Federal Energy Regulatory Commission. *FERC Directs PJM to Create New Rules for Co-Located Generation and Load.* Federal Energy Regulatory Commission. Dec. 18, 2025.

[91] Texas Legislature. *Senate Bill 6, 89th Legislature.* Texas Legislature. June 20, 2025.

[92] Mike Jacobs. *Connection Costs: Loophole Costs Customers Over $4 Billion to Connect Data Centers to Power Grid.* Union of Concerned Scientists, Sept. 29, 2025.

[93] Michigan Attorney General. *AG Nessel on the MPSC Approving DTE's Application to Service Saline Data Center.* Michigan Attorney General. Dec. 18, 2025.

[94] Sydney Forrester et al. *Retail Electricity Price and Cost Trends: 2024 Update.* Lawrence Berkeley National Laboratory, Dec. 2024.

[95] Ethan Howland. "PJM capacity prices set another record with 22% jump". In: *Utility Dive* (July 23, 2025).

[96] Ethan Howland. *Data centers 'primary reason' for high PJM capacity prices: market monitor.* Utility Dive. Oct. 2, 2025.

[97] Federal Energy Regulatory Commission. *Order No. 2023: Improvements to Generator Interconnection Procedures and Agreements.* Federal Energy Regulatory Commission, July 28, 2023.

[98] White & Case LLP. *DOE directs FERC to accelerate interconnection of data centers.* White & Case LLP. Oct. 27, 2025.

[99] Constellation Energy. *Constellation to Launch Crane Clean Energy Center, Restoring Jobs and Carbon-Free Power to The Grid.* Constellation Energy. Sept. 20, 2024.

[100] GE Vernova. *Homer City Redevelopment and Kiewit*

announce country's largest natural gas-powered data center campus to support AI and HPC demand. GE Vernova. Apr. 2, 2025.

[101] U.S. Energy Information Administration. *Electric Power Monthly*. U.S. Energy Information Administration. 2025.

[102] Dave Jones et al. *US Electricity 2025 – Special Report*. Ember, Mar. 12, 2025.

[103] Michael J. Bommarito. *Gigawatt-Scale Datacenter Projects: The New AI Infrastructure Paradigm*. 2025.

[104] Zachary Skidmore. "Homer City and Kiewit unveil plans for 4.5GW natural gas powered AI data center in Pennsylvania". In: *Data Center Dynamics* (Apr. 2, 2025).

[105] Goldman Sachs Research. *GS SUSTAIN: Generational Growth: AI, data centers and the coming US power demand surge*. Goldman Sachs, Apr. 29, 2024.

[106] Evan Halper. "AI's hunger for electric power is threatening U.S. climate goals". In: *The Washington Post* (Nov. 19, 2024).

[107] Uptime Institute. *Uptime Institute Global Data Center Survey 2024*. Uptime Institute, July 2024.

[108] Caiwei Chen. "China built hundreds of AI data centers to catch the AI boom. Now many stand unused." In: *MIT Technology Review* (Mar. 26, 2025).

[109] Google. *2024 Environmental Report*. Google LLC, July 2024.

[110] Shannon Heckt. "SCC approves Chesterfield gas plant and Dominion rate hike, creates new rate class for data centers". In: *Virginia Mercury* (Nov. 25, 2025).

[111] Abbe Ramanan. "How data centers have emerged as a threat to state efforts to combat the climate crisis". In: *Utility Dive* (July 25, 2025).

[112] Stephen P. Holland, Matthew J. Kotchen, and Andrew J. Yates. "Why marginal CO2 emissions are not decreasing for US electricity: Estimates and implications for climate policy". In: *Proceedings of the National Academy of Sciences* 119.8 (Feb. 14, 2022), e2116632119.

[113] Michael Terrell. *New nuclear clean energy agreement with Kairos Power*. Google. Oct. 14, 2024.

[114] Amazon Staff. *Amazon signs agreements for innovative nuclear energy projects to address growing energy demands*. Amazon.com. Oct. 16, 2024.

[115] U.S. Department of Energy. *Energy Department Closes Loan to Restart Nuclear Power Plant in Pennsylvania*. U.S. Department of Energy. Nov. 18, 2025.

[116] Constellation Energy. *Constellation, Meta Sign 20-Year Deal for Clean, Reliable Nuclear Energy in Illinois*. Constellation Energy. June 3, 2025.

[117] Kairos Power. *Nuclear Regulatory Commission Approves Construction Permit for Hermes Demonstration Reactor.* Kairos Power. Dec. 12, 2023.

[118] Google. *Our first advanced nuclear reactor project with Kairos Power and Tennessee Valley Authority.* Google. Aug. 18, 2025.

[119] Amazon Staff. *How Amazon is helping to build one of the first modular nuclear reactor facilities in the United States.* Amazon.com. Oct. 16, 2025.

[120] Switch. *Oklo and Switch Form Landmark Strategic Relationship to Deploy 12 Gigawatts of Advanced Nuclear Power.* Switch. Dec. 18, 2024.

[121] U.S. Energy Information Administration. *Plant Vogtle Unit 4 begins commercial operation.* U.S. Energy Information Administration. May 1, 2024.

[122] World Nuclear News. "US Regulator Completes First SMR Design Certification Rulemaking". In: *World Nuclear News* (Jan. 23, 2023).

[123] NuScale Power. *Utah Associated Municipal Power Systems (UAMPS) and NuScale Power Agree to Terminate the Carbon Free Power Project (CFPP).* NuScale Power. Nov. 8, 2023.

[124] Google. *Google and Fervo launch first-of-its-kind geothermal project.* Google. Nov. 28, 2023.

[125] Zanskar. *Zanskar raises $115M Series C Following Record-Setting Year of Geothermal Discoveries.* Zanskar. Jan. 21, 2026.

[126] Robin Whitlock. "Renewables provided 26 percent of US electricity in 2025 according to SUN DAY campaign". In: *Renewable Energy Magazine* (Dec. 29, 2025).

[127] Lazard. *Lazard's Levelized Cost of Energy Analysis—Version 17.0.* Lazard, June 2024.

[128] Michigan Legislature. *Public Act 181 of 2024 (Senate Bill 237): Enterprise Data Center Tax Exemptions.* Michigan Legislature. Dec. 30, 2024.

[129] Congressional Research Service. *The Section 45Q Tax Credit for Carbon Sequestration.* Congressional Research Service, June 18, 2024.

[130] Langdon Winner. "Do Artifacts Have Politics?" In: *Daedalus* 109.1 (1980), pp. 121–136.

[131] USDA National Agricultural Statistics Service. *Land Values 2024 Summary.* United States Department of Agriculture, Aug. 2024.

[132] Northeast-Midwest Institute. *Benefits and Impacts of Brownfields Development.* Northeast-Midwest Institute, July 2008.

[133] *Internal Revenue Code Section 198: Expensing of Environmental Remediation Costs.* 1997.

[134] 117th U.S. Congress. *Inflation Reduction Act of 2022*. 2022.

[135] Port KC. "$100 Billion Data Center Development to bring Infrastructure, Investment and Jobs to the Northland". In: *Port KC Press Release* (Aug. 26, 2025).

[136] Associated Press. "Amazon to spend $20B on data centers in Pennsylvania as AI fuels demand". In: *Spotlight PA* (June 9, 2025).

[137] Brian Allnutt. "$1 billion data center plan hits roadblock in Howell Township". In: *Planet Detroit* (Sept. 25, 2025).

[138] Jackie Smith. "Rezoning request for Meta-backed data center near Howell withdrawn, officials say". In: *MLive* (Dec. 8, 2025).

[139] Brian Allnutt. "Can 957 signatures halt a $1 billion data center in Augusta Township?" In: *Planet Detroit* (Aug. 29, 2025).

[140] Jason Ma. "Opposition leads developer to withdraw data center proposal in Kalkaska County, Michigan". In: *Data Center Dynamics* (Nov. 25, 2025).

[141] Brian Allnutt. "Ypsilanti Township residents speak out against University of Michigan's 'biggest, baddest' data center". In: *Planet Detroit* (June 20, 2025).

[142] Zachary Skidmore. *DTE Energy signs 1.4GW power deal with unnamed hyperscaler*. Oct. 30, 2025.

[143] Gregory T. Lyman. *How Much Water Does Golf Use and Where Does It Come From?* GCSAA for United States Golf Association Water Summit. 2012.

[144] Molly Taft. "How Much Water Do AI Data Centers Really Use?" In: *Undark Magazine (via WIRED)* (Dec. 16, 2025).

[145] Sachi Kitajima Mulkey. "The surging demand for data is guzzling Virginia's water". In: *Grist* (May 8, 2024).

[146] Brad Smith. *Microsoft will replenish more water than it consumes by 2030*. Microsoft Corporation. Sept. 21, 2020.

[147] Great Lakes-St. Lawrence River Basin Water Resources Council. *Great Lakes-St. Lawrence River Basin Water Resources Compact*. 2008.

[148] Kasia Tarczynska. *Money Lost to the Cloud: How Data Centers Benefit from State and Local Government Subsidies*. Good Jobs First, Oct. 2016.

[149] Blackstone. *Blackstone Announces Agreement to Acquire AirTrunk in a A$24B Transaction*. Blackstone. Sept. 4, 2024.

[150] Blackstone. *Blackstone Reports Third-Quarter 2024 Earnings*. Blackstone. Oct. 17, 2024.

[151] Stephen Schwarzman. *Blackstone Second Quarter 2024 Earnings Call Transcript*. Blackstone. July 18, 2024.

[152] Blackstone. *Blackstone Funds Complete Acquisition of QTS Realty Trust.* Blackstone. Aug. 31, 2021.

[153] Sebastian Moss. "Blackstone's prospective data center pipeline hits $100bn, on top of $70bn portfolio". In: *Data Center Dynamics* (Oct. 24, 2024).

[154] Federal Highway Administration. *Highway Statistics 2023: Interstate Highway System Historical Costs.* U.S. Department of Transportation, 2024.

[155] The Planetary Society. "How Much Did the Apollo Program Cost?" In: *The Planetary Society* (2024).

[156] Congressional Budget Office. *CBO's Estimates of ARRA's Economic Impact.* Congressional Budget Office, Feb. 2012.

[157] Deepa Seetharaman, Krystal Hu, and Arsheeya Bajwa. "OpenAI Discussed Government Loan Guarantees for Chip Plants, Not Data Centers, Altman Says". In: *U.S. News & World Report* (Nov. 6, 2025).

[158] KKR. *KKR and GIP Complete Acquisition of CyrusOne.* KKR. Mar. 25, 2022.

[159] DigitalBridge. *DigitalBridge and IFM Investors Complete $11 Billion Take-Private of Switch.* DigitalBridge. Dec. 6, 2022.

[160] Compass Datacenters. *Brookfield Infrastructure and Ontario Teachers' to Acquire Compass Datacenters from RedBird Capital Partners and Azrieli Group.* Compass Datacenters. June 20, 2023.

[161] Amazon.com, Inc. *2024 Annual Report (Form 10-K).* U.S. Securities and Exchange Commission. Feb. 7, 2025.

[162] Alphabet Inc. *2024 Annual Report (Form 10-K).* U.S. Securities and Exchange Commission. Feb. 4, 2025.

[163] Meta Platforms, Inc. *Fourth Quarter 2024 Results Conference Call.* Meta Platforms. Jan. 29, 2025.

[164] Kif Leswing. "Nvidia's $100 billion OpenAI deal showcases chipmaker's growing investment portfolio". In: *CNBC* (Sept. 26, 2025).

[165] Lance Murray. "North Texas' Aligned Data Centers To Be Acquired for $40 Billion by BlackRock, Nvidia, xAI, Microsoft, and Others". In: *Dallas Innovates* (Oct. 15, 2025).

[166] Dan Swinhoe. "BlackRock to acquire Global Infrastructure Partners for $3 billion cash + shares". In: *Data Center Dynamics* (Jan. 12, 2024).

[167] Equinix, Inc. "Equinix 2024 Annual Report (Form 10-K)". In: *SEC Filing* (Feb. 12, 2025).

[168] BlackRock. *BlackRock, Global Infrastructure Partners, Microsoft, and MGX Launch New AI Partnership to Invest in Data Centers and Supporting Power*

Infrastructure. BlackRock. Sept. 17, 2024.

[169] Kif Leswing. "Nvidia-backed GPU cloud provider CoreWeave surges to $19 billion valuation". In: *CNBC* (May 1, 2024).

[170] Jordan Novet. "Nvidia-backed CoreWeave closes flat at $40 after biggest U.S. tech IPO since 2021". In: *CNBC* (Mar. 28, 2025).

[171] Mike Jacobs. *Data Centers Are Already Increasing Your Energy Bills*. Union of Concerned Scientists, Sept. 29, 2025.

[172] Ethan Howland. "FERC urged to reject AEP waiver request for PJM capacity sale". In: *Utility Dive* (Dec. 2, 2025).

[173] Charles Paullin. "Virginia Regulators Approve New Dominion Rates, Assign More Costs to Data Centers". In: *Inside Climate News* (Jan. 7, 2026).

[174] Justin Stevens and Eric Brooks. *Adapting Utility Tariffs for Data Center Driven Load Growth*. ScottMadden Energy Practice, Dec. 19, 2025.

[175] Josh Saul et al. "How AI Data Centers Are Sending Your Power Bill Soaring". In: *Bloomberg* (Sept. 29, 2025).

[176] Tom Rutigliano. *Building Data Centers Without Breaking PJM*. Natural Resources Defense Council, Sept. 30, 2025.

[177] Government Finance Officers Association. "Creation, Implementation, and Evaluation of Tax Increment Financing". In: *GFOA Best Practices* (Feb. 28, 2014).

[178] Paige Tortorelli, Pippa Stevens, and Agne Tolockaite. "Tax breaks for tech giants' data centers mean less income for states". In: *CNBC* (June 20, 2025).

[179] Jesse Colombo. *The Late 1990s Telecom Bubble*. 2024.

[180] Doug O'Laughlin. "Lessons from History: The Rise and Fall of the Telecom Bubble". In: *Fabricated Knowledge* (Oct. 16, 2023).

[181] Waldemar Szlezak et al. "Beyond the Bubble: Why We Think AI Infrastructure Will Compound Long after the Hype". In: *KKR Global Insights* (Nov. 2025).

[182] Joint Legislative Audit and Review Commission. *Data Centers in Virginia*. Commonwealth of Virginia, Dec. 9, 2024.

[183] Louisiana Economic Development. *Meta Selects Northeast Louisiana as Site of $10 Billion Artificial Intelligence Optimized Data Center*. Louisiana Economic Development. Dec. 4, 2024.

[184] Erin Socha. "Lawsuit delays $12B data center in Kansas City as community, environmental group voice concerns". In: *Kansas Reflector* (Nov. 7, 2025).

[185] Dan Swinhoe. "Related Digital's 1GW Michigan data center project

to be part of OpenAI's Stargate". In: *Data Center Dynamics* (Oct. 31, 2025).

[186] Sheila Jasanoff. *The Ethics of Invention: Technology and the Human Future.* New York: W. W. Norton, 2016.

[187] Beth LeBlanc. "Michigan's Benson has familial, professional ties to company pushing data center project in Saline". In: *Michigan Advance* (Nov. 12, 2025).

[188] David Zin. *Film Incentives in Michigan.* Michigan Senate Fiscal Agency, Sept. 2010.

[189] Doug Ringler. *Michigan Economic Growth Authority Tax Credit Program: Performance Audit Report.* Michigan Office of the Auditor General, Sept. 2017.

[190] Pew Charitable Trusts. "Faulty Forecasts: Michigan's MEGA Tax Credit". In: *Pew Charitable Trusts Research and Analysis* (Dec. 2, 2015).

[191] David Chernicoff. "Texas Senate Bill 6: A Bellwether On How States May Approach Data Center Energy Use". In: *Data Center Frontier* (July 2, 2025).

[192] Monica Samayoa. "Oregon Legislature passes 'POWER Act,' targeting industrial energy users like data centers". In: *Oregon Public Broadcasting* (June 5, 2025).

[193] 117th U.S. Congress. *CHIPS and Science Act.* Aug. 9, 2022.

[194] TSMC. *TSMC Arizona and U.S. Department of Commerce Announce up to US$6.6 Billion in Proposed CHIPS Act Direct Funding.* TSMC. Apr. 8, 2024.

[195] Alissa Widman Neese, Tyler Buchanan, and Jessica Boehm. "How Intel's $20 billion Ohio One project is reshaping the Columbus region". In: *Axios Columbus* (Jan. 22, 2024).

[196] Donald J. Trump. *Remarks on Artificial Intelligence Infrastructure Development and an Exchange With Reporters.* The American Presidency Project. Jan. 21, 2025.

[197] Ashish Vaswani et al. "Attention Is All You Need". In: *Advances in Neural Information Processing Systems.* Vol. 30. 2017.

[198] Kasper Groes Albin Ludvigsen. "The carbon footprint of GPT-4". In: *Towards Data Science* (July 18, 2023).

[199] OpenAI and SoftBank. *Announcing The Stargate Project.* OpenAI. Jan. 21, 2025.

[200] State Council of the People's Republic of China. *New Generation Artificial Intelligence Development Plan.* State Council of China. July 20, 2017.

[201] DeepSeek. *About DeepSeek.* 2023.

[202] Jing Liu. *"East Data, West Compute" Project: Weaving a Nationwide Computing Network [Dong Shu Xi Suan].* chinese. Xinhua News Agency. Apr. 20, 2022.

[203] Jensen Huang. *NVIDIA's Jensen Huang on Securing American Leadership on AI.* Center for Strategic and International Studies. Dec. 3, 2025.

[204] Jake Sullivan. *Remarks by National Security Advisor Jake Sullivan at the Special Competitive Studies Project Global Emerging Technologies Summit.* The White House. Sept. 16, 2022.

[205] U.S. Department of Commerce, Bureau of Industry and Security. *Commerce Strengthens Restrictions on Advanced Computing Semiconductors, Semiconductor Manufacturing Equipment, and Supercomputing Items to Countries of Concern.* U.S. Department of Commerce. Oct. 17, 2023.

[206] House Foreign Affairs Committee. *Chairman Mast, HFAC, Advances AI Overwatch Act.* Jan. 21, 2026.

[207] OpenAI. *Stargate Community.* Jan. 20, 2026.

[208] United States District Court, Northern District of California. *Musk v. Altman et al.* Jan. 2026.

[209] Maureen Farrell and Rob Copeland. "Saudi Arabia Plans $40 Billion Push Into Artificial Intelligence". In: *The New York Times* (Mar. 19, 2024).

[210] Peter Judge. "Singapore lifts data center moratorium - but sets conditions". In: *Data Center Dynamics* (Jan. 12, 2022).

[211] Ashley Capoot. "Anthropic, Google, OpenAI and xAI granted up to $200 million for AI work from Defense Department". In: *CNBC* (July 14, 2025).

[212] Thomas Novelly. "The Air Force wants to put private AI data centers on its bases, raising security, land-use fears". In: *Defense One* (Oct. 24, 2025).

[213] European Union. *Regulation (EU) 2024/1689 of the European Parliament and of the Council (Artificial Intelligence Act).* European Union. July 12, 2024.

[214] The White House. *Executive Order 14110: Safe, Secure, and Trustworthy Development and Use of Artificial Intelligence.* The White House. Oct. 30, 2023.

[215] CBRE. *2024 U.S. Real Estate Market Outlook: Data Centers.* CBRE. 2024.

[216] EdgeCore Digital Infrastructure. *EdgeCore Digital Infrastructure Announces Development of Reno, Nevada Data Center Campus.* EdgeCore Digital Infrastructure. Aug. 10, 2023.

[217] Compass Datacenters. *Beyond Stick-Built: The Manufacturing Advantage in Data Center Construction.* Compass Datacenters. Apr. 10, 2025.

[218] Vapor IO. *Kinetic Grid Platform.* Vapor IO. 2024.

[219] Apple Machine Learning Research. *Introducing Apple's On-Device and*

Server Foundation Models. Apple. June 10, 2024.

[220] Apple Security Research. *Private Cloud Compute: A new frontier for AI privacy in the cloud.* Apple. June 10, 2024.

[221] Caren Chang, Chengji Yan, and Taj Darra. *On-device GenAI APIs as part of ML Kit help you easily build with Gemini Nano.* Google Android Developers Blog. May 2025.

[222] Ken Addison. *Qualcomm Snapdragon X2 Elite Architectural Details and Performance Preview.* Signal65. Nov. 20, 2025.

[223] vLLM Project. *vLLM: A high-throughput and memory-efficient inference and serving engine for LLMs.* PyTorch Foundation. May 2025.

[224] Fatih E. Nar et al. "Why vLLM Is the Best Choice for AI Inference Today". In: *Red Hat Developer Blog* (Oct. 30, 2025).

[225] Global Market Insights. *Enterprise LLM Market Size and Share Report, 2025-2034.* Global Market Insights, Sept. 2025.

[226] John Roach. *Microsoft finds underwater datacenters are reliable, practical and use energy sustainably.* Microsoft. Sept. 14, 2020.

[227] SpaceX. *Falcon 9.* SpaceX. 2024.

[228] Rebecca Szkutak. "200 VCs wanted to get into Lumen Orbit's $11M

seed round". In: *TechCrunch* (Dec. 11, 2024).

[229] Constellation Energy. *Constellation to Launch Crane Clean Energy Center, Restoring Jobs and Carbon-Free Power to The Grid.* Constellation Energy. Sept. 20, 2024.

[230] Stockholm Data Parks. *Stockholm Data Parks: Making the Modern Sustainable City.* Stockholm Data Parks. 2024.

[231] Edged Energy. *Edged Energy Launches in the U.S. with Four Ultra-Efficient, AI Ready Data Centers Delivering 300+ MW and Saving 1.2 Billion Gallons of Water Each Year.* Edged Energy. Feb. 13, 2024.

[232] CyrusOne. *Chandler, AZ: PHX1-PHX8.* CyrusOne. 2024.

[233] Aligned Data Centers. *Advanced Data Center Cooling.* Aligned Data Centers. 2024.

[234] Oracle Corporation. *Planet - Environmental and Social Impact.* Oracle Corporation. 2024.

[235] Federal Energy Regulatory Commission. *Order No. 1920: Building for the Future Through Electric Regional Transmission Planning and Cost Allocation.* Federal Energy Regulatory Commission, May 13, 2024.

[236] Central Statistics Office (Ireland). *Data Centres Metered Electricity Consumption 2024.* Central Statistics Office. 2025.

[237] U.S. Department of Energy. *DOE Releases New Report Evaluating Increase in Electricity Demand from Data Centers*. U.S. Department of Energy. Dec. 20, 2024.

[238] World Economic Forum. *The Future of Jobs Report 2025*. World Economic Forum, Jan. 2025.

[239] Alex Rough. "Data Centers on the 21st Century Battlefield". In: *War on the Rocks* (Sept. 5, 2025).

www.ingramcontent.com/pod-product-compliance
Lightning Source LLC
Chambersburg PA
CBHW052119270326
41930CB00012B/2680